SpringerBriefs in Electrical and Computer Engineering

Computational Electromagnetics

Series editor

Rakesh Mohan Jha, Bangalore, India

More information about this series at http://www.springer.com/series/13885

Hema Singh · H.L. Sneha · Rakesh Mohan Jha

Scattering Cross Section of Unequal Length Dipole Arrays

 Springer

Hema Singh
Centre for Electromagnetics
CSIR-National Aerospace Laboratories
Bangalore, Karnataka
India

Rakesh Mohan Jha
Centre for Electromagnetics
CSIR-National Aerospace Laboratories
Bangalore, Karnataka
India

H.L. Sneha
Centre for Electromagnetics
CSIR-National Aerospace Laboratories
Bangalore, Karnataka
India

ISSN 2191-8112   ISSN 2191-8120   (electronic)
SpringerBriefs in Electrical and Computer Engineering
ISSN 2365-6239   ISSN 2365-6247   (electronic)
SpringerBriefs in Computational Electromagnetics
ISBN 978-981-287-789-5   ISBN 978-981-287-790-1   (eBook)
DOI 10.1007/978-981-287-790-1

Library of Congress Control Number: 2015947420

Springer Singapore Heidelberg New York Dordrecht London
© The Author(s) 2016
This work is subject to copyright. All rights are reserved by the Publisher, whether the whole or part
of the material is concerned, specifically the rights of translation, reprinting, reuse of illustrations,
recitation, broadcasting, reproduction on microfilms or in any other physical way, and transmission
or information storage and retrieval, electronic adaptation, computer software, or by similar or dissimilar
methodology now known or hereafter developed.
The use of general descriptive names, registered names, trademarks, service marks, etc. in this
publication does not imply, even in the absence of a specific statement, that such names are exempt from
the relevant protective laws and regulations and therefore free for general use.
The publisher, the authors and the editors are safe to assume that the advice and information in this
book are believed to be true and accurate at the date of publication. Neither the publisher nor the
authors or the editors give a warranty, express or implied, with respect to the material contained herein or
for any errors or omissions that may have been made.

Printed on acid-free paper

Springer Science+Business Media Singapore Pte Ltd. is part of Springer Science+Business Media
(www.springer.com)

To Professor R. Narasimha

In Memory of Dr. Rakesh Mohan Jha
Great scientist, mentor, and excellent
human being

Dr. Rakesh Mohan Jha was a brilliant contributor to science, a wonderful human being, and a great mentor and friend to all of us associated with this book. With a heavy heart we mourn his sudden and untimely demise and dedicate this book to his memory.

Preface

The radar cross section (RCS) of phased array depends on the design parameters. The dimension and geometric configuration of dipole elements are the most important factors that control the array performance including radiation behavior. This book presents the RCS estimation of an array with unequal length dipoles. The signal reflections within the antenna system and the mutual coupling effect are considered to arrive at the total RCS for series and parallel feed. The analytical description includes the dependence of RCS of dipole array on design parameters, viz., dipole length, interelement spacing, geometrical and electrical properties of couplers, and terminal impedances. It is shown that the antenna design parameters like dipole length, geometric configuration, and terminal impedance can be optimized toward the RCS control of phased array. The theoretical formulation and illustrations in this book provide an insight to the reader regarding the role of design parameters of dipole antenna element in overall array RCS.

Hema Singh
H. L. Sneha
Rakesh Mohan Jha

Acknowledgments

We would like to thank Mr. Shyam Chetty, Director, CSIR-National Aerospace Laboratories, Bangalore for his permission and support to write this SpringerBrief.

We would also like to acknowledge valuable suggestions from our colleagues at the Centre for Electromagnetics, Dr. R.U. Nair, Dr. Shiv Narayan, Dr. Balamati Choudhury, and Mr. K.S. Venu during the course of writing this book. We express our sincere thanks to Mr. Harish S. Rawat, Ms. Neethu P.S., Mr. Umesh V. Sharma, and Mr. Bala Ankaiah, the project staff at the Centre for Electromagnetics, for their consistent support during the preparation of this book.

But for the concerted support and encouragement from Springer, especially the efforts of Suvira Srivastav, Associate Director, and Swati Meherishi, Senior Editor, Applied Sciences & Engineering, it would not have been possible to bring out this book within such a short span of time. We very much appreciate the continued support by Ms. Kamiya Khatter and Ms. Aparajita Singh of Springer toward bringing out this brief.

Contents

About the Authors

Dr. Hema Singh is currently working as Senior Scientist in Centre for Electromagnetics of CSIR-National Aerospace Laboratories, Bangalore, India. Earlier, she was Lecturer in EEE, BITS, Pilani, India during 2001–2004. She obtained her Ph.D. degree in Electronics Engineering from IIT-BHU, Varanasi India in 2000. Her active area of research is Computational Electromagnetics for Aerospace Applications. More specifically, the topics she has contributed to, are GTD/UTD, EM analysis of propagation in an indoor environment, phased arrays, conformal antennas, radar cross section (RCS) studies including Active RCS Reduction. She received the Best Woman Scientist Award in CSIR-NAL, Bangalore for period of 2007–2008 for her contribution in the areas of phased antenna array, adaptive arrays, and active RCS reduction. Dr. Singh has co-authored one book, one book chapter, and over 120 scientific research papers and technical reports.

Ms. H.L. Sneha obtained her BE (ECE) degree from Visvesvaraya Technological University, Karnataka. She was a Project Engineer at the Centre for Electromagnetics of CSIR-National Aerospace Laboratories, Bangalore, where she worked on radar cross-section studies, phased arrays, and mutual coupling effects in dipole arrays.

Dr. Rakesh Mohan Jha was Chief Scientist & Head, Centre for Electromagnetics, CSIR-National Aerospace Laboratories, Bangalore. Dr. Jha obtained a dual degree in BE (Hons.) EEE and M.Sc. (Hons.) Physics from BITS, Pilani (Raj.) India, in 1982. He obtained his Ph.D. (Engg.) degree from Department of Aerospace Engineering of Indian Institute of Science, Bangalore in 1989, in the area of computational electromagnetics for aerospace applications. Dr. Jha was a SERC (UK) Visiting Post-Doctoral Research Fellow at University of Oxford, Department of Engineering Science in 1991. He worked as an Alexander von Humboldt Fellow at the Institute for High-Frequency Techniques and Electronics of the University of

Karlsruhe, Germany (1992–1993, 1997). He was awarded the Sir C.V. Raman Award for Aerospace Engineering for the Year 1999. Dr. Jha was elected Fellow of INAE in 2010, for his contributions to the EM Applications to Aerospace Engineering. He was also the Fellow of IETE and Distinguished Fellow of ICCES. Dr. Jha has authored or co-authored several books, and more than five hundred scientific research papers and technical reports. He passed away during the production of this book of a cardiac arrest.

List of Figures

List of Tables

Scattering Cross Section of Unequal Length Dipole Arrays

Abstract The antenna radar cross section (RCS) depends on the field scattered by the antenna toward the receiver. It has two components, viz., structural RCS and antenna mode RCS. The latter component dominates over the former, especially if the antenna is mounted on a low-observable platform. The reduction in the scattering due to the presence of antennas on the surface is one of the concerns toward stealth technology. In order to achieve this objective, a detailed and accurate analysis of antenna mode scattering is required. In practical phased array, one cannot ignore the finite dimensions of antenna elements, coupling effect, and the role of feed network while estimating the antenna RCS. This book presents the RCS estimation of an array with unequal-length dipoles. The signal reflections within the antenna system and the mutual coupling effect are considered to arrive at the total RCS for series and parallel feed. The scattering due to higher order reflections is neglected. The computations are valid for any arbitrary array configurations, including side-by-side arrangement, parallel-in-echelon, etc.

1 Introduction

The scattering cross section of an antenna depends on the field scattered by the antenna toward the receiver. The antenna radar cross section (RCS) consists of both structural RCS and antenna mode RCS. However the latter component dominates over the former, especially if the antenna is mounted on a low observable platform (Zhanget al. 2010). The reduction in the scattering due to the presence of antennas on the surface is one of the primary demands toward stealth technology. In order to achieve this objective, a detailed and accurate analysis of antenna mode scattering is required. The parameters that affect the scattering within the antenna system such as architecture of feed network, impedance matching, and mutual coupling (Liu and You 2011; Niow et al. 2011) have to be studied.

The total scattered field of an antenna array comprises of the fields reflected from different impedance mismatches at each level of feed network. The corresponding

© The Author(s) 2016

H. Singh et al., *Scattering Cross Section of Unequal Length Dipole Arrays*,
SpringerBriefs in Computational Electromagnetics,
DOI 10.1007/978-981-287-790-1_1

scattered field magnitudes can be expressed in terms of reflection coefficients and terminal impedances of the array elements.

Further, the values of the terminal impedances are influenced by the mutual coupling effect, which depends on the array geometry (Jenn 1995). This makes necessary to analyze the signal path through each of the junction and mismatches within the antenna system toward the RCS estimation of phased arrays.

The scattering behavior of phased array has been analyzed by researchers using different techniques. A finite dipole array has been studied in view of its radiation and scattering characteristics with the compensation of coupling effect (Liao et al. 2006a, b). The dipoles of equal length (half-wavelength) are considered. The moment method along with RWG basis functions are used to calculate scattered field in the presence of mutual coupling. Although the parametric analysis of scattering behavior of dipole array is presented, the feed network is not considered. The numerical technique like FDTD method is also used to calculate the scattering from impedance-loaded dipole array (Zengrui and Junhong 2006; Zengrui et al. 2007) without feed network. The scattering behavior of antenna array with feed network is presented for infinitesimal dipole array (Jenn and Lee 1995; Jenn and Flokas 1996). However, the reported results ignore the mutual coupling effect within the dipole elements. In practical phased array, one cannot ignore the finite dimensions of antenna elements, coupling effect, and the role of feed network while estimating the antenna RCS.

The study on RCS of finite dipole array with series and parallel feed network including mutual coupling effect has been reported in (Sneha et al. 2012a, b, 2013). The dipoles were considered to be of equal length in different geometric configurations i.e., side-by-side, collinear, and parallel-in-echelon. In this book, the RCS estimation of unequal length dipole array is described. The computed results are presented for both series and parallel feed network. Figure 1 shows a uniform linear array of center-fed dipoles each with length of $2l_n$, $n = 1, 2,...N$. The interelement spacing is taken as d. The distance of the dipoles from the reference plane is assumed to be h_n. The geometry of dipole array considered is generalized and it can be converted to any standard array configuration, like side-by-side or parallel-in-echelon through appropriate changes in the design parameters.

The array design considered here differs from that reported by Sneha et al. (2012a, b) in two aspects. At first, it uses an extra component, which can be either the waveguide bend or transmission line, in order to connect the dipole terminals to the corresponding phase-shifter. This extra component is assumed to be perfectly matched to the phase-shifters so as to have zero contribution toward total array RCS. Otherwise, the impedance mismatch within this extra line is necessary to be included in calculations of scattered field. Second, the dipole lengths are taken to be unequal as an attempt toward the reduction of scattering within the antenna system.

Section 2 describes the formulation for the total scattered field of unequal length finite dipole array for both series and parallel feed networks. The mutual coupling effect is included in the estimation of impedance at the dipole terminals. Section 3 presents the simulation results for antenna RCS varying different design parameters. The observations and inferences are summarized in Sect. 4.

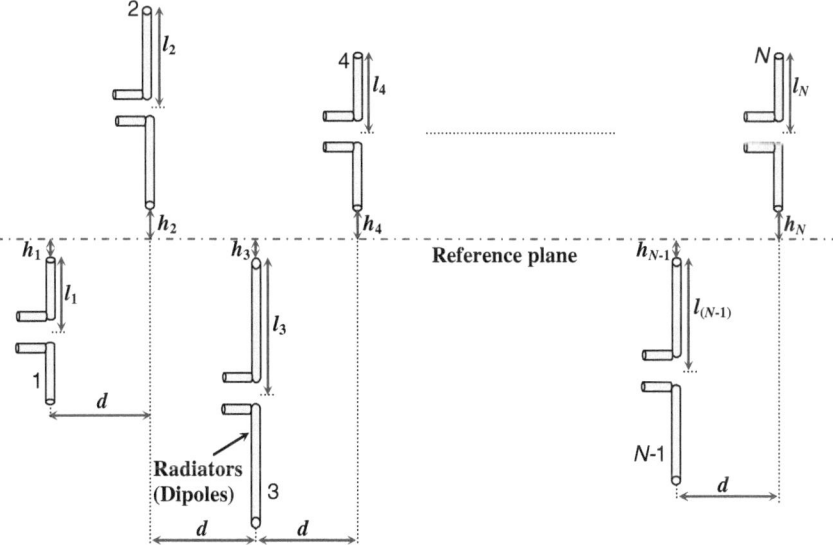

Fig. 1 Typical unequal length dipole array

2 Formulation for the RCS of Dipole Array

The scattered field of a lossless x-polarized nth dipole with cosine distributed surface current is given by (Sneha et al. 2012a)

$$\vec{E}_n^s(\theta, \phi) = \left[\frac{j\eta_0}{4\lambda Z_{s_n}} \left(\int_{\Delta l_n} \cos(kl_n)\, dl_n \right)^2 (\cos\theta)\, \vec{E}_n^r(\theta, \phi) \right] \frac{e^{-jk\vec{R}}}{R}\hat{x} \qquad (1)$$

where λ is the wavelength, η_0 is the impedance of free space, k is the wave number, $\vec{k}$ is the wave vector, R is the distance between the target and the observation point, l_n is the length of nth dipole element, $\vec{E}_n^r(\theta, \phi)$ is the total reflected field toward the aperture, and Z_{s_n} is the impedance of nth dipole element, expressed in terms of its resistance, R_{s_n} and reactance, X_{s_n} as (Balanis 2005)

$$Z_{s_n} = R_{s_n} + jX_{s_n} \qquad (2)$$

where the resistance and reactance are expressed in terms of cosine and sine integrals

$$R_{s_n} = \frac{\eta}{2\pi} \left[\begin{array}{l} C + \ln(kl_n) - C_i(kl_n) + \frac{1}{2}\sin(kl_n)\{S_i(2kl_n) - 2S_i(kl_n)\} \\ + \frac{1}{2}\cos(kl_n)\left\{C + \ln\left(kl_n/2\right) + C_i(2kl_n) - 2C_i(kl_n)\right\} \end{array} \right] \qquad (3)$$

$$X_{S_n} = \frac{\eta}{4\pi} \left[\begin{array}{l} 2S_i(kl_n) + \cos(kl_n)\{2S_i(kl_n) - S_i(2kl_n)\} \\ - \sin(kl_n) \left\{ 2C_i(kl_n) - C_i(2kl_n) - C_i\left(\frac{2ka_n^2}{l_n}\right) \right\} \end{array} \right] \tag{4}$$

Here $C_i(kl_n)$ and $S_i(kl_n)$ are the cosine and sine integrals and (l_n, a_n) indicate the length and radius of nth dipole in the phased array.

In order to determine the total RCS of N-element dipole array, scattered fields at each of its elements are summed-up as follows:

$$\sigma(\theta, \phi) = \frac{4\pi}{\lambda^2} \left| \sum_{n=1}^{N} \left\{ \frac{j\eta_0}{4\lambda Z_{S_n}} \left(\int_{\Delta l_n} \cos(kl_n) \, dl_n \right)^2 (\cos\theta) \, \vec{E}_n^r(\theta, \phi) \right\} \right|^2 \tag{5}$$

The total scattered field, and hence the RCS of the dipole array, can be decomposed in terms of individual reflections at each level including antenna aperture and the feed network. These reflected fields are determined by the magnitude of reflection coefficients and the path through which the signal propagates. As a result, the RCS of the dipole array is specific to the structure of the feed network used to excite the elements.

In this section, the scattering from unequal length dipole array for two different feed networks, (i) series feed (Fig. 2) and (ii) parallel feed (Fig. 3) is formulated.

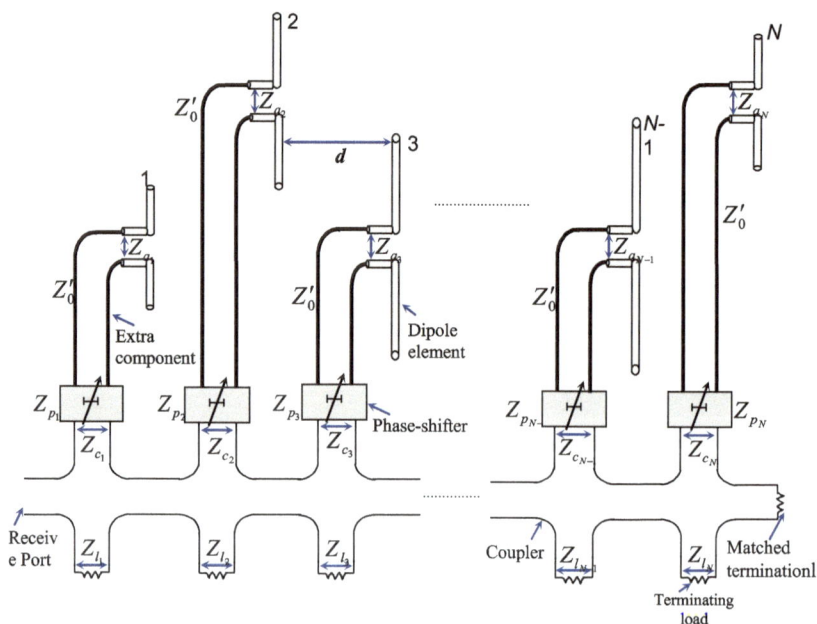

Fig. 2 Impedances at various levels in series-fed unequal length dipole array

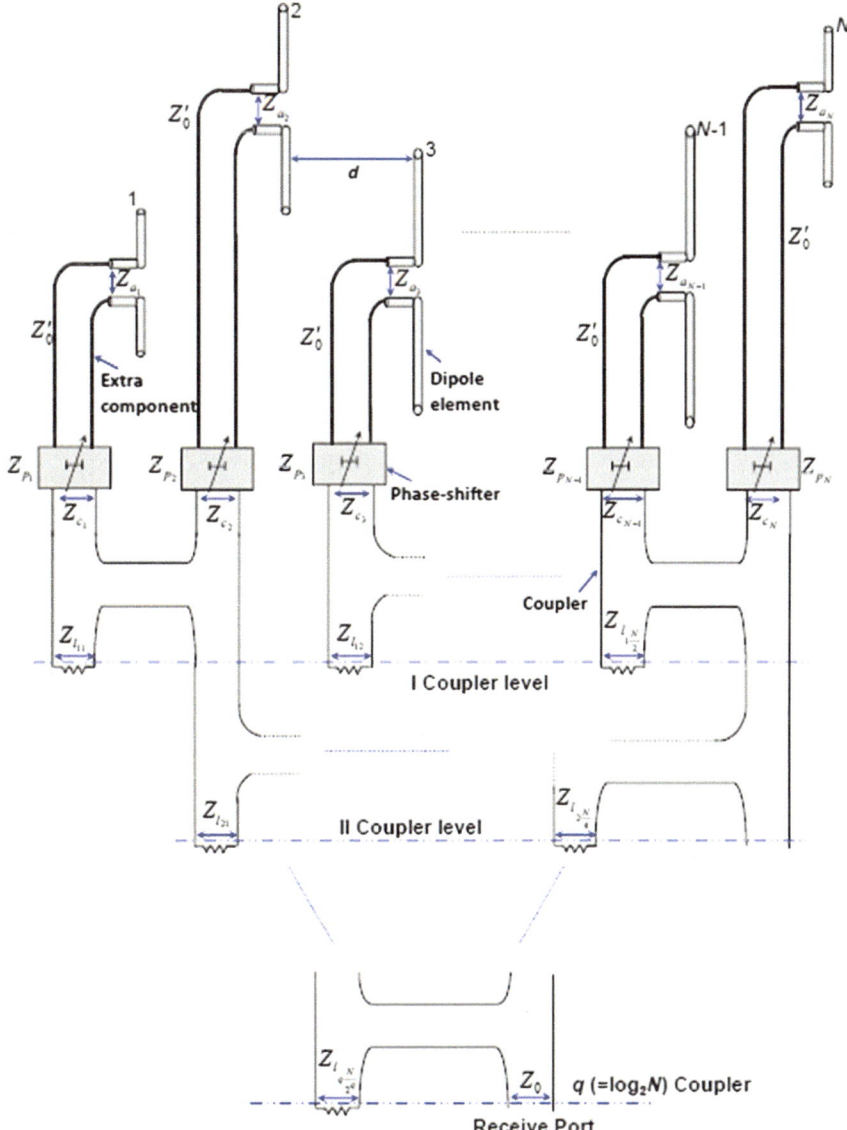

Fig. 3 Impedances at various levels in parallel-fed unequal length dipole array

It can be seen that the architecture of feed network remains same for either type of feeds till one reaches the couplers. It is inferred that the contribution to the total RCS remains identical for all the levels preceding the level of coupler(s) in antenna system.

2.1 Reflection at the Dipole Terminals

The first source of scattering for a signal impinging on the array (Fig. 4) is the junction of radiators and connectors. The corresponding RCS equation is given as

$$\sigma_r(\theta, \phi) = \sum_{n=1}^{N} \left\{ \frac{j\eta_0}{4\lambda Z_{S_n}} \left(\int_{\Delta l_n} \cos(kl_n) \, dl_n \right)^2 (\cos\theta) \, r_{r_n} e^{j2(n-1)\alpha} \right\} \tag{6}$$

where α is the interelement space delay of incident wave along array axis and r_{r_n} is the reflection coefficient of nth dipole.

The magnitude of this reflection coefficient is determined by the mismatch between the impedances of dipoles and the impedance of the additional line (Fig. 4) connecting the dipole terminals to the inputs of phase-shifters,

$$r_{r_n} = \left| \frac{Z_{a_n} - Z_0'}{Z_{a_n} + Z_0'} \right| \tag{7}$$

Here Z_0' is the impedance of the extra component (assumed to be same as the impedance of phase-shifter Z_0) and Z_{a_n} is the impedance at the nth dipole terminal. Mathematically (Balanis 2005),

$$Z_{a_n} = \sum_{y=1}^{N} z_{a_{x,y}} \frac{I_y}{I_x}; \; z_{a_{x,y}} = \begin{pmatrix} z_{a_{1,1}} & z_{a_{1,2}} & \cdots & z_{a_{1,N}} \\ z_{a_{2,1}} & z_{a_{2,2}} & \cdots & z_{a_{2,N}} \\ \vdots & \vdots & \ddots & \vdots \\ z_{a_{N,1}} & z_{a_{N,2}} & \cdots & z_{a_{N,N}} \end{pmatrix} \tag{8}$$

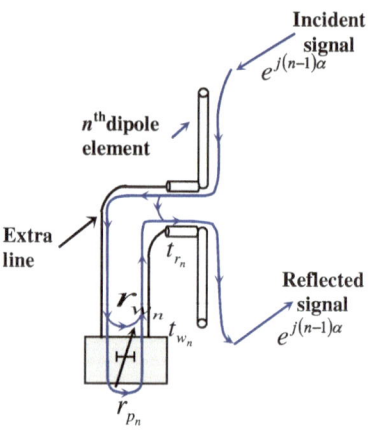

Fig. 4 Reflection and transmission coefficients till phase-shifters

where I_n indicates the current fed to nth dipole at its terminals, according to the aperture distribution, and $Z_{a_{x,y}}$ indicates the impedance matrix of array elements. The impedance matrix of a phased array comprises of both self and mutual impedances of its array elements.

For a dipole array with infinitesimally thin, parallel, center-fed, unequal length dipoles, the self impedance is computed using (3) and (4). The expression for the mutual impedance of a dipole array is given by

$$Z_{\text{mutual}_{x,y}} = R_{\text{mutual}_{x,y}} + jX_{\text{mutual}_{x,y}} \tag{9}$$

where (King 1957)

$$R_{\text{mutual}_{x,y}} = 15 \begin{bmatrix} \cos k(l_x - h)\{C_i(u_o) + C_i(v_o) - C_i(u_1) - C_i(v_1)\} \\ + \sin k(l_x - h)\{-S_i(u_o) + S_i(v_o) + S_i(u_1) - S_i(v_1)\} \\ + \cos k(l_x + h)\{C_i(u_0') + C_i(v_0') - C_i(u_2) - C_i(v_2)\} \\ + \sin k(l_x + h)\{-S_i(u_0') + S_i(v_0') + S_i(u_2) - S_i(v_2)\} \\ + \cos k(l_x - 2l_y - h)\times \\ \{-C_i(u_1) - C_i(v_1) + C_i(u_3) + C_i(v_3)\} \\ + \sin k(l_x - 2l_y - h)\begin{Bmatrix} S_i(u_1) - S_i(v_1) \\ - S_i(u_3) + S_i(v_3) \end{Bmatrix} \\ + \cos k(l_x + 2l_y + h)\begin{Bmatrix} - C_i(u_2) - C_i(v_2) \\ + C_i(u_4) + C_i(v_4) \end{Bmatrix} \\ + \sin k(l_x + 2l_y + h)\begin{Bmatrix} S_i(u_2) - S_i(v_2) \\ - S_i(u_4) + S_i(v_4) \end{Bmatrix} \\ + 2\cos kl_x \cos kh \begin{Bmatrix} -C_i(w_1) - C_i(y_1) \\ + C_i(w_2) + C_i(y_2) \end{Bmatrix} \\ + 2\cos kl_x \sin kh \begin{Bmatrix} S_i(w_1) - S_i(y_1) \\ - S_i(w_2) + S_i(y_2) \end{Bmatrix} \\ + 2\cos kl_x \cos k(2l_y + h)\begin{Bmatrix} C_i(w_2) + C_i(y_2) \\ - C_i(w_3) - C_i(y_3) \end{Bmatrix} \\ + 2\cos kl_x \sin k(2l_y + h)\begin{Bmatrix} -S_i(w_2) + S_i(y_2) \\ + S_i(w_3) - S_i(y_3) \end{Bmatrix} \end{bmatrix} \tag{10}$$

$$
X_{\text{mutual}_{x,y}} = 15 \left[
\begin{aligned}
& \cos k(l_x - h) \left\{ \begin{aligned} & -S_i(u_0) - S_i(v_0) \\ & +S_i(u_1) + S_i(v_1) \end{aligned} \right\} \\
& + \sin k(l_x - h) \left\{ \begin{aligned} & -C_i(u_0) + C_i(v_0) \\ & +C_i(u_1) - C_i(v_1) \end{aligned} \right\} \\
& + \cos k(l_x + h) \left\{ \begin{aligned} & -S_i(u_0') - S_i(v_0') \\ & +S_i(u_2) + S_i(v_2) \end{aligned} \right\} \\
& + \sin k(l_x + h) \left\{ \begin{aligned} & -C_i(u_0') + C_i(v_0') \\ & +C_i(u_2) - C_i(v_2) \end{aligned} \right\} \\
& + \cos k(l_x - 2l_y - h) \left\{ \begin{aligned} & S_i(u_1) + S_i(v_1) \\ & -S_i(u_3) - S_i(v_3) \end{aligned} \right\} \\
& + \sin k(l_x - 2l_y - h) \left\{ \begin{aligned} & C_i(u_1) - C_i(v_1) \\ & -C_i(u_3) + C_i(v_3) \end{aligned} \right\} \\
& + \cos k(l_x + 2l_y + h) \left\{ \begin{aligned} & S_i(u_2) + S_i(v_2) \\ & -S_i(u_4) - S_i(v_4) \end{aligned} \right\} \\
& + \sin k(l_x + 2l_y + h) \left\{ \begin{aligned} & C_i(u_2) - C_i(v_2) \\ & -C_i(u_4) + C_i(v_4) \end{aligned} \right\} \\
& + 2 \cos kl_x \cos kh \left\{ \begin{aligned} & S_i(w_1) + S_i(y_1) \\ & -S_i(w_2) - S_i(y_2) \end{aligned} \right\} \\
& + 2 \cos kl_x \sin kh \left\{ \begin{aligned} & C_i(w_1) - C_i(y_1) \\ & -C_i(w_2) + C_i(y_2) \end{aligned} \right\} \\
& + 2 \cos kl_x \cos k(2l_y + h) \left\{ \begin{aligned} & -S_i(w_2) - S_i(y_2) \\ & +S_i(w_3) + S_i(y_3) \end{aligned} \right\} \\
& + 2 \cos kl_x \sin k(2l_y + h) \left\{ \begin{aligned} & -C_i(w_2) + C_i(y_2) \\ & +C_i(w_3) - C_i(y_3) \end{aligned} \right\}
\end{aligned}
\right] \tag{11}
$$

where

$$
u_0 = k \left\{ \sqrt{d_r^2 + (h - l_x)^2} + (h - l_x) \right\} \tag{12a}
$$

$$
v_0 = k \left\{ \sqrt{d_r^2 + (h - l_x)^2} - (h - l_x) \right\} \tag{12b}
$$

$$
u_0' = k \left\{ \sqrt{d_r^2 + (h + l_x)^2} - (h + l_x) \right\} \tag{12c}
$$

$$v'_0 = k \left\{ \sqrt{d_r^2 + (h + l_x)^2} + (h + l_x) \right\} \tag{12d}$$

$$u_1 = k \left\{ \sqrt{d_r^2 + (h - l_x + l_y)^2} + (h - l_x + l_y) \right\} \tag{12e}$$

$$v_1 = k \left\{ \sqrt{d_r^2 + (h - l_x + l_y)^2} - (h - l_x + l_y) \right\} \tag{12f}$$

$$u_2 = k \left\{ \sqrt{d_r^2 + (h + l_x + l_y)^2} - (h + l_x + l_y) \right\} \tag{12g}$$

$$v_2 = k \left\{ \sqrt{d_r^2 + (h + l_x + l_y)^2} + (h + l_x + l_y) \right\} \tag{12h}$$

$$u_3 = k \left\{ \sqrt{d_r^2 + (h - l_x + 2l_y)^2} + (h - l_x + 2l_y) \right\} \tag{12i}$$

$$v_3 = k \left\{ \sqrt{d_r^2 + (h - l_x + 2l_y)^2} - (h - l_x + 2l_y) \right\} \tag{12j}$$

$$u_4 = k \left\{ \sqrt{d_r^2 + (h + l_x + 2l_y)^2} - (h + l_x + 2l_y) \right\} \tag{12k}$$

$$v_4 = k \left\{ \sqrt{d_r^2 + (h + l_x + 2l_y)^2} + (h + l_x + 2l_y) \right\} \tag{12l}$$

$$w_1 = k \left\{ \sqrt{d_r^2 + h^2} - h \right\} \tag{12m}$$

$$y_1 = k \left\{ \sqrt{d_r^2 + h^2} + h \right\} \tag{12n}$$

$$w_2 = k \left\{ \sqrt{d_r^2 + (h + l_y)^2} - (h + l_y) \right\} \tag{12o}$$

$$y_2 = k \left\{ \sqrt{d_r^2 + (h + l_y)^2} + (h + l_y) \right\} \tag{12p}$$

$$w_3 = k \left\{ \sqrt{d_r^2 + (h + 2l_y)^2} - (h + 2l_y) \right\} \tag{12q}$$

$$y_3 = k \left\{ \sqrt{d_r^2 + (h + 2l_y)^2} + (h + 2l_y) \right\} \tag{12r}$$

Here (l_x, l_y) indicate the half-lengths of xth and yth dipoles, respectively; d_r represents the relative distance between the dipoles in xth and yth array positions; h represents the staggered height w.r.t. the reference plane. The expressions (10) through (12a) hold good for an array with parallel-dipoles with any geometrical configuration, like equal-length, unequal-length, side-by-side, parallel-in-echelon etc. However, this formulation does not hold for skewed dipole arrays.

2.2   Reflection at the Terminals of Extra Component

Next, the signal would travel toward the input port of phase-shifters via the extra line. As this section of feed is assumed to be perfectly matched, its reflection coefficient r_{w_n} will be zero. Thus the contribution of reflections occurring at the terminals of the extra line to the array RCS is nil. This is done for the sake of convenience in the RCS estimation of dipole array. Thus,

$$\sigma_w(\theta, \phi) = 0 \qquad (13)$$

2.3   Reflection at the Phase-Shifters

The signal reaching the phase-shifters will get reflected at its input ports due to the impedance mismatch. The corresponding reflection coefficient r_{p_n} is given as (Sneha et al. 2013)

$$r_{p_n} = \left| \frac{Z_{p_n} - Z_0'}{Z_{p_n} + Z_0'} \right| (Z_0' = Z_0 \text{ for matched extra line}) \qquad (14)$$

where Z_{p_n} is the impedance at the end terminals of the delay lines. This yields the RCS due to the reflections at the terminals of phase-shifters as

$$\sigma_p(\theta, \phi) = \sum_{n=1}^{N} \left\{ \frac{j\eta_0}{4\lambda Z_{s_n}} \left(\int_{\Delta l_n} \cos(kl_n) \, dl_n \right)^2 (\cos \theta) \, t_{r_n}^2 t_{w_n}^2 r_{p_n} e^{j2(n-1)\alpha} \right\} \qquad (15)$$

where t_{r_n} is the transmission coefficient of nth dipole, r_{p_n} is the reflection coefficient of nth phase-shifter and t_{w_n} is the transmission coefficient of nth extra line. However for the matched extra line, t_{w_n} is always one,

$$t_{w_n} = \sqrt{1 - r_{w_n}^2}; \; r_{w_n} = 0 \qquad (16)$$

2.4 Reflection at the Coupler Ports

Next level in the feed network which contributes to the RCS is the junction of phase-shifters and the input ports of couplers (Figs. 2 and 3).

2.4.1 Signal Reflection at the Input Port(s) of the Couplers Connected to Phase-Shifters

The contribution of reflections at the input terminal of couplers connected to phase-shifters for the RCS of dipole array is given as

$$\sigma_c(\theta, \phi) = \sum_{n=1}^{N} \left\{ \frac{j\eta_0}{4\lambda Z_{s_n}} \left(\int\limits_{\Delta l_n} \cos(kl_n)\, dl_n \right)^2 (\cos \theta)\, t_{r_n}^2 t_{w_n}^2 t_{p_n}^2 r_{c_n} e^{j2(n-1)\zeta} \right\} \quad (17)$$

where t_{p_n} is the transmission coefficient of nth phase-shifter, r_{c_n} is the reflection coefficient of coupler port connected to the end of phase-shifter (Sneha et al. 2012b) and $\zeta = \alpha + \alpha_s$; α_s is the interelement phase to scan antenna beam along the array axis. However, the calculation for the reflection coefficients at this level depends on the type of feed network. This is because, the couplers in parallel feed network interact with multiple antenna elements, unlike in series feed (Sneha et al. 2012a).

2.4.2 Signal Reflection at Other Coupler Ports

The signal that enters through the input port of coupler, gets reflected within the coupler. Since the geometrical arrangement and the nature of couplers differ for the series and parallel feed network, the RCS formulation also varies.

For series feed network: The RCS due to the couplers is expressed as

$$\sigma_s(\theta, \phi) = \sum_{n=1}^{N} \left\{ \begin{array}{l} \frac{j\eta_0}{4\lambda Z_{s_n}} \left(\int\limits_{\Delta l_n} \cos(kl_n)\, dl_n \right)^2 (\cos \theta)\, \vec{E}_{1_n}^r(\theta, \phi) \\[2ex] + \frac{j\eta_0}{4\lambda Z_{s_n}} \left(\int\limits_{\Delta l_n} \cos(kl_n)\, dl_n \right)^2 (\cos \theta)\, \vec{E}_{2_n}^r(\theta, \phi) \\[2ex] + \frac{j\eta_0}{4\lambda Z_{s_n}} \left(\int\limits_{\Delta l_n} \cos(kl_n)\, dl_n \right)^2 (\cos \theta)\, \vec{E}_{3_n}^r(\theta, \phi) \\[2ex] + \frac{j\eta_0}{4\lambda Z_{s_n}} \left(\int\limits_{\Delta l_n} \cos(kl_n)\, dl_n \right)^2 (\cos \theta)\, \vec{E}_{4_n}^r(\theta, \phi) \end{array} \right\} \quad (18)$$

$$\sigma_s(\theta, \phi) = \sum_{n=1}^{N} \left[\frac{j\eta_0}{4\lambda Z_{s_n}} \left(\int_{\Delta l_n} \cos(kl_n) dl_n \right)^2 (\cos\theta) \left\{ \begin{array}{l} \vec{E}_{1_n}^r(\theta, \phi) + \vec{E}_{2_n}^r(\theta, \phi) \\ + \vec{E}_{3_n}^r(\theta, \phi) + \vec{E}_{4_n}^r(\theta, \phi) \end{array} \right\} \right] \quad (19)$$

where $\vec{E}_{1_n}^r(\theta, \phi)$ is the reflected field at nth dipole arising from the signal propagating toward the next antenna element, given by

$$\vec{E}_{1_n}^r(\theta, \varphi) = \left[t_{r_n} t_{w_n} t_{p_n} r_{l_n} j c_n e^{j(n-1)\zeta} \sum_{m=n+1}^{N} \left(t_{r_m} t_{w_m} t_{p_m} j c_m e^{j(m-1)\zeta} \prod_{i=n}^{m-1} t_{c_i} e^{j\psi} \right) \right] \quad (20)$$

$\vec{E}_{2_n}^r(\theta, \phi)$ is the reflected field at nth dipole arising from the signal propagating toward previous antenna element(s) in the array, given by

$$\vec{E}_{2_n}^r(\theta, \phi) = \left[t_{r_n} t_{w_n} t_{p_n} j c_n e^{j(n-1)\zeta} \sum_{m=1}^{n-1} t_{r_m} t_{w_m} t_{p_m} r_{l_m} j c_m e^{j(m-1)\zeta} \prod_{i=m}^{n-1} t_{c_i} e^{j\psi} \right] \quad (21)$$

$\vec{E}_{3_n}^r(\theta, \phi)$ is the reflected field at nth dipole due to signal propagating toward the terminating load and is expressed as

$$\vec{E}_{3_n}^r(\theta, \phi) = r_{l_n} t_{r_n}^2 t_{w_n}^2 t_{p_n}^2 t_{c_n}^2 e^{j2(n-1)\zeta} \quad (22)$$

and $\vec{E}_{4_n}^r(\theta, \phi)$ is the reflected field at nth dipole due to the signal propagating toward the receive port, given by

$$\vec{E}_{4_n}^r(\theta, \phi) = \left[r_{in} t_{r_n}^2 t_{w_n}^2 t_{p_n}^2 (jc_n)^2 e^{j2(n-1)\zeta} \left(\prod_{i=1}^{n-1} t_{c_i} e^{j\psi} \right)^2 \right] \quad (23)$$

Figure 5 illustrates the signal paths and the corresponding reflected fields at nth dipole in a series-fed dipole array. The signal path is shown in blue color.

For parallel feed network: The RCS due to the coupler level(s) is expressed as

$$\sigma_{sd_1}(\theta, \phi) = \sum_{n=1,3\ldots}^{N-1} \left\{ \begin{array}{l} \frac{j\eta_0}{4\lambda Z_{sn}} \left(\int_{\Delta l_n} \cos(kl_n) dl_n \right)^2 (\cos\theta) \vec{E}_{n_1}^r(\theta, \phi) \\ + \frac{j\eta_0}{4\lambda Z_{s_{(n+1)}}} \left(\int_{\Delta l_{(n+1)}} \cos(kl_{(n+1)}) dl_{(n+1)} \right)^2 (\cos\theta) \vec{E}_{(n+1)_1}^r(\theta, \phi) \end{array} \right\}$$

$$(24)$$

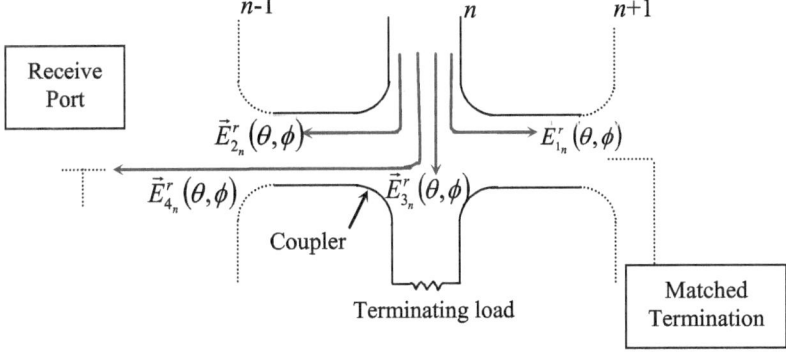

Fig. 5 Signal paths and the corresponding reflected fields at nth dipole element

due to first coupler level
where

$$
\vec{E}^r_{n_1}(\theta,\phi) = t_{r_n} t_{w_n} t_{p_n} e^{j(n-1)\zeta} \left\{ r_{s_{1i}} c_{1i} e^{j\psi} \left(\begin{array}{c} c_{1i} e^{j\psi} t_{r_n} t_{w_n} t_{p_n} e^{j(n-1)\zeta} \\ + t_{r_{n+1}} t_{w_{n+1}} t_{p_{n+1}} e^{jn\zeta} t_{c_{1i}} \end{array} \right) + r_{d_{1i}} t_{c_{1i}} \left(\begin{array}{c} t_{r_n} t_{w_n} t_{p_n} e^{j(n-1)\zeta} t_{c_{1i}} \\ + t_{r_{n+1}} t_{w_{n+1}} t_{p_{n+1}} e^{jn\zeta} c_{1i} e^{j\psi} \end{array} \right) \right\} \tag{25a}
$$

$$
\vec{E}^r_{(n+1)_1}(\theta,\phi) = t_{r_{n+1}} t_{w_{n+1}} t_{p_{n+1}} e^{jn\zeta} \left\{ r_{s_{1i}} t_{c_{1i}} \left(\begin{array}{c} t_{r_n} t_{w_n} t_{p_n} e^{j(n-1)\zeta} c_{1i} e^{j\psi} \\ + t_{c_{1i}} t_{r_{n+1}} t_{w_{n+1}} t_{p_{n+1}} e^{jn\zeta} \end{array} \right) + r_{d_{1i}} c_{1i} e^{j\psi} \left(\begin{array}{c} t_{r_n} t_{w_n} t_{p_n} e^{j(n-1)\zeta} t_{c_{1i}} \\ + t_{r_{n+1}} t_{w_{n+1}} t_{p_{n+1}} e^{jn\zeta} c_{1i} e^{j\psi} \end{array} \right) \right\} \tag{25b}
$$

and

$$
\sigma_{sd_2}(\theta,\phi) = \sum_{n=1,5\ldots}^{N-3} \left\{ \begin{array}{c} \frac{j\eta_0}{4\lambda Z_{s_n}} \left(\int_{\Delta l_n} \cos(kl_n)\, dl_n \right)^2 (\cos\theta)\, \vec{E}^r_{n_2}(\theta,\phi) \\ \frac{j\eta_0}{4\lambda Z_{s(n+1)}} \left(\int_{\Delta l_{(n+1)}} \cos(kl_{(n+1)})\, dl_{(n+1)} \right)^2 (\cos\theta)\, \vec{E}^r_{(n+1)_2}(\theta,\phi) \\ + \frac{j\eta_0}{4\lambda Z_{s(n+2)}} \left(\int_{\Delta l_{(n+2)}} \cos(kl_{(n+2)})\, dl_{(n+2)} \right)^2 (\cos\theta)\, \vec{E}^r_{(n+2)_2}(\theta,\phi) \\ + \frac{j\eta_0}{4\lambda Z_{s(n+3)}} \left(\int_{\Delta l_{(n+3)}} \cos(kl_{(n+3)})\, dl_n \right)^2 (\cos\theta)\, \vec{E}^r_{(n+3)_2}(\theta,\phi) \end{array} \right\} \tag{26}
$$

due to second coupler level
where

$$
\vec{E}^r_{n_2}(\theta,\phi) = t_{r_n} t_{w_n} t_{p_n} e^{j(n-1)\zeta} c_{1i} e^{j\psi} t_{s_{1i}}
\left[
r_{s_{2i'}} c_{2i'} e^{j\psi}
\left\{
\begin{array}{l}
t_{r_n} t_{w_n} t_{p_n} e^{j(n-1)\zeta} c_{1i} e^{j\psi} \\[4pt]
\times\, t_{s_{1i}} c_{2i'} e^{j\psi} \\[4pt]
+\, t_{r_{n+1}} t_{w_{n+1}} t_{p_{n+1}} e^{jn\zeta} \\[4pt]
\times\, t_{c_{1i}} t_{s_{1i}} c_{2i'} e^{j\psi} \\[4pt]
+\, t_{r_{n+2}} t_{w_{n+2}} t_{p_{n+2}} e^{j(n+1)\zeta} \\[4pt]
\times\, c_{1(i+1)} e^{j\psi} t_{s_{1(i+1)}} t_{c_{2i'}} \\[4pt]
+\, t_{r_{n+3}} t_{w_{n+3}} t_{p_{n+3}} e^{j(n+2)\zeta} \\[4pt]
\times\, t_{c_{1(i+1)}} t_{s_{1(i+1)}} t_{c_{2i'}}
\end{array}
\right\}
\right.
$$
$$
\left.
+\, r_{d_{2i'}} t_{c_{2i'}}
\left\{
\begin{array}{l}
t_{r_n} t_{w_n} t_{p_n} e^{j(n-1)\zeta} c_{1i} e^{j\psi} t_{s_{1i}} t_{c_{2i'}} \\[4pt]
+\, t_{r_{n+1}} t_{w_{n+1}} t_{p_{n+1}} e^{jn\zeta} t_{c_{1i}} t_{s_{1i}} t_{c_{2i'}} \\[4pt]
+\, t_{r_{n+2}} t_{w_{n+2}} t_{p_{n+2}} e^{j(n+1)\zeta} \\[4pt]
\times\, c_{1(i+1)} e^{j\psi} t_{s_{1(i+1)}} c_{2i'} e^{j\psi} \\[4pt]
+\, t_{r_{n+3}} t_{w_{n+3}} t_{p_{n+3}} e^{j(n+2)\zeta} \\[4pt]
\times\, t_{c_{1(i+1)}} t_{s_{1(i+1)}} c_{2i'} e^{j\psi}
\end{array}
\right\}
\right]
\tag{27a}
$$

$$
\vec{E}^r_{(n+1)_2}(\theta,\phi) = t_{r_{n+1}} t_{w_{n+1}} t_{p_{n+1}} e^{jn\zeta} t_{c_{1i}} t_{s_{1i}}
\left[
r_{s_{2i'}} c_{2i'} e^{j\psi}
\left\{
\begin{array}{l}
t_{r_n} t_{w_n} t_{p_n} e^{j(n-1)\zeta} c_{1i} e^{j\psi} \\[4pt]
\times\, t_{s_{1i}} c_{2i'} e^{j\psi} +\, t_{r_{n+1}} t_{w_{n+1}} \\[4pt]
\times\, t_{p_{n+1}} e^{jn\zeta} t_{c_{1i}} \\[4pt]
\times\, t_{s_{1i}} c_{2i'} e^{j\psi} +\, t_{r_{n+2}} t_{w_{n+2}} \\[4pt]
\times\, t_{p_{n+2}} e^{j(n+1)\zeta} c_{1(i+1)} e^{j\psi} \\[4pt]
\times\, t_{s_{1(i+1)}} t_{c_{2i'}} +\, t_{r_{n+3}} t_{w_{n+3}} \\[4pt]
\times\, t_{p_{n+3}} e^{j(n+2)\zeta} t_{c_{1(i+1)}} t_{s_{1(i+1)}} t_{c_{2i'}}
\end{array}
\right\}
\right.
$$
$$
\left.
+\, r_{d_{2i'}} t_{c_{2i'}}
\left\{
\begin{array}{l}
t_{r_n} t_{w_n} t_{p_n} e^{j(n-1)\zeta} c_{1i} e^{j\psi} \\[4pt]
\times\, t_{s_{1i}} t_{c_{2i'}} +\, t_{r_{n+1}} t_{w_{n+1}} \\[4pt]
\times\, t_{p_{n+1}} e^{jn\zeta} t_{c_{1i}} t_{s_{1i}} t_{c_{2i'}} +\, t_{r_{n+2}} t_{w_{n+2}} t_{p_{n+2}} \\[4pt]
\times\, e^{j(n+1)\zeta} c_{1(i+1)} e^{j\psi} t_{s_{1(i+1)}} c_{2i'} e^{j\psi} \\[4pt]
+\, t_{r_{n+3}} t_{w_{n+3}} t_{p_{n+3}} e^{j(n+2)\zeta} t_{c_{1(i+1)}} t_{s_{1(i+1)}} c_{2i'} e^{j\psi}
\end{array}
\right\}
\right]
\tag{27b}
$$

$$\vec{E}^r_{(n+2)_2}(\theta, \phi) = t_{r_{n+2}} t_{w_{n+2}} t_{p_{n+2}} e^{j(n+1)\zeta} c_{1(i+1)} e^{j\psi} t_{s_{1(i+1)}}$$

$$\times \left[r_{s_{2i'}} t_{c_{2i'}} \left\{ \begin{array}{l} t_{r_n} t_{w_n} t_{p_n} e^{j(n-1)\zeta} c_{1i} e^{j\psi} t_{s_{1i}} c_{2i'} e^{j\psi} \\[4pt] + t_{r_{n+1}} t_{w_{n+1}} t_{p_{n+1}} e^{jn\zeta} t_{c_{1i}} t_{s_{1i}} c_{2i'} e^{j\psi} \\[4pt] + t_{r_{n+2}} t_{w_{n+2}} t_{p_{n+2}} e^{j(n+1)\zeta} \\[4pt] \times e^{j\psi} c_{1(i+1)} t_{s_{1(i+1)}} t_{c_{2i'}} \\[4pt] + t_{r_{n+3}} t_{w_{n+3}} t_{p_{n+3}} e^{j(n+2)\zeta} \\[4pt] \times t_{c_{1(i+1)}} t_{s_{1(i+1)}} t_{c_{2i'}} \end{array} \right\} \right.$$

$$\left. + r_{d_{2i'}} c_{2i'} e^{j\psi} \left\{ \begin{array}{l} t_{r_n} t_{w_n} t_{p_n} e^{j(n-1)\zeta} c_{1i} e^{j\psi} t_{s_{1i}} t_{c_{2i'}} \\[4pt] + t_{r_{n+1}} t_{w_{n+1}} t_{p_{n+1}} e^{jn\zeta} t_{c_{1i}} t_{s_{1i}} t_{c_{2i'}} \\[4pt] + t_{r_{n+2}} t_{w_{n+2}} t_{p_{n+2}} e^{j(n+1)\zeta} e^{j\psi} \\[4pt] \times c_{1(i+1)} t_{s_{1(i+1)}} c_{2i'} e^{j\psi} \\[4pt] + t_{r_{n+3}} t_{p_{n+3}} e^{j(n+2)\zeta} t_{c_{1(i+1)}} t_{s_{1(i+1)}} c_{2i'} e^{j\psi} \end{array} \right\} \right] \tag{27c}$$

$$\vec{E}^r_{(n+3)_2}(\theta, \phi) = t_{r_{n+3}} t_{w_{n+3}} t_{p_{n+3}} e^{j(n+2)\zeta} t_{c_{1(i+1)}} t_{s_{1(i+1)}}$$

$$\times \left[r_{s_{2i'}} t_{c_{2i'}} \left\{ \begin{array}{l} t_{r_n} t_{w_n} t_{p_n} e^{j(n-1)\zeta} c_{1i} e^{j\psi} t_{s_{1i}} c_{2i'} e^{j\psi} + t_{r_{n+1}} t_{w_{n+1}} t_{p_{n+1}} e^{jn\zeta} \\[4pt] \times t_{c_{1i}} t_{s_{1i}} c_{2i'} e^{j\psi} + t_{r_{n+2}} t_{w_{n+2}} t_{p_{n+2}} e^{j(n+1)\zeta} c_{1(i+1)} e^{j\psi} \\[4pt] \times t_{s_{1(i+1)}} t_{c_{2i'}} + t_{r_{n+3}} t_{w_{n+3}} t_{p_{n+3}} e^{j(n+2)\zeta} t_{c_{1(i+1)}} t_{s_{1(i+1)}} t_{c_{2i'}} \end{array} \right\} \right.$$

$$\left. + r_{d_{2i'}} c_{2i'} e^{j\psi} \left\{ \begin{array}{l} t_{r_n} t_{w_n} t_{p_n} e^{j(n-1)\zeta} c_{1i} e^{j\psi} t_{s_{1i}} t_{c_{2i'}} + t_{r_{n+1}} t_{w_{n+1}} t_{p_{n+1}} e^{jn\zeta} \\[4pt] \times t_{c_{1i}} t_{s_{1i}} t_{c_{2i'}} + t_{r_{n+2}} t_{w_{n+2}} t_{p_{n+2}} e^{j(n+1)\zeta} c_{1(i+1)} e^{j\psi} t_{s_{1(i+1)}} \\[4pt] \times c_{2i'} e^{j\psi} + t_{r_{n+3}} t_{w_{n+3}} t_{p_{n+3}} e^{j(n+2)\zeta} t_{c_{1(i+1)}} t_{s_{1(i+1)}} c_{2i'} e^{j\psi} \end{array} \right\} \right] \tag{27d}$$

It is noted that the mutual impedance is included in the calculation of coefficients (c, t_c, t_s) of the coupler level in the feed network.

The total RCS of unequal length dipole array with mutual coupling effect is expressed as

$$\sigma(\theta, \phi) = \frac{4\pi}{\lambda^2} \left\{ |\sigma_r(\theta, \phi)|^2 + |\sigma_p(\theta, \phi)|^2 + |\sigma_c(\theta, \phi)|^2 + |\sigma_s(\theta, \phi)|^2 \right\} \tag{28}$$

(for series feed network)

$$\sigma(\theta, \phi) = \frac{4\pi}{\lambda^2} \left\{ |\sigma_r(\theta, \phi)|^2 + |\sigma_p(\theta, \phi)|^2 + |\sigma_c(\theta, \phi)|^2 + |\sigma_{sd_1}(\theta, \phi)|^2 \right\} \qquad (29)$$

(for parallel feed network till first level of couplers)

$$\sigma(\theta, \phi) = \frac{4\pi}{\lambda^2} \left\{ |\sigma_r(\theta, \phi)|^2 + |\sigma_p(\theta, \phi)|^2 + |\sigma_c(\theta, \phi)|^2 + |\sigma_{sd_1}(\theta, \phi)|^2 + |\sigma_{sd_2}(\theta, \phi)|^2 \right\}$$

$$(30)$$

(for parallel feed network till second level of couplers).

3   Results and Discussion

This book presents a study on the RCS estimation of unequal length, uniformly spaced linear dipole array in the presence of mutual coupling. The extra component, connecting antenna element with the phase shifter in the feed network is assumed to be perfectly matched for all the cases. The computed results are discussed for the dipole arrays with series and parallel feed networks in two separate sub-sections. The scattered field is calculated by tracing the impinging signal path through the components of phased array system.

3.1   RCS Estimation of Series-Fed Dipole Array

In this subsection, the RCS pattern is computed based on the formulation described above.

3.1.1   Equal-Length Dipole Array

Figure 6 shows both the broadside ($\theta_s = 0°$) and scanned ($\theta_s = 35°$) RCS of the array, with mutual coupling effect. The geometry of the dipole array considered is shown in Fig. 7.

It is a 30-element array consisting of infinitesimally thin, equal length $\lambda/2$ dipoles spaced at 0.25λ. The staggered height of all the dipoles is considered to be $-\lambda/4$ w.r. t. the plane of reference. The characteristic impedance of the delay-line and terminating load impedance are taken as 125 and 235 Ω, respectively. The excitation is uniform unit amplitude distribution.

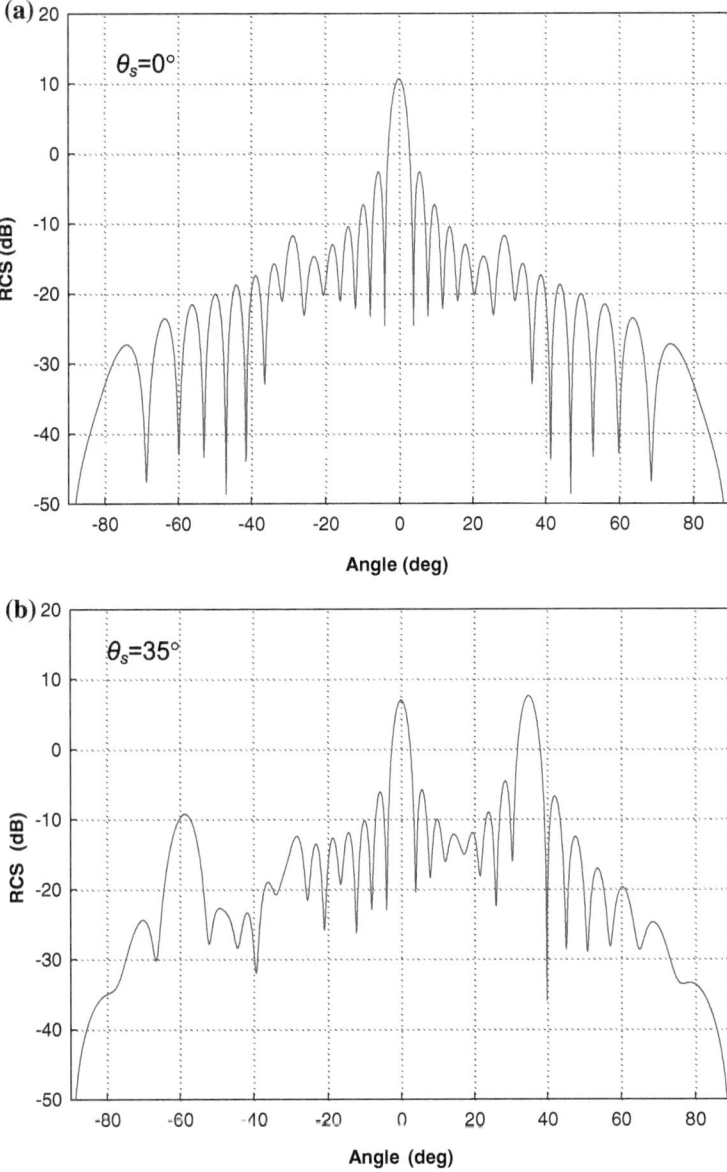

Fig. 6 RCS of an equal length linear dipole array in the presence of mutual coupling. **a** $\theta_s = 0°$.
b $\theta_s = 35°$

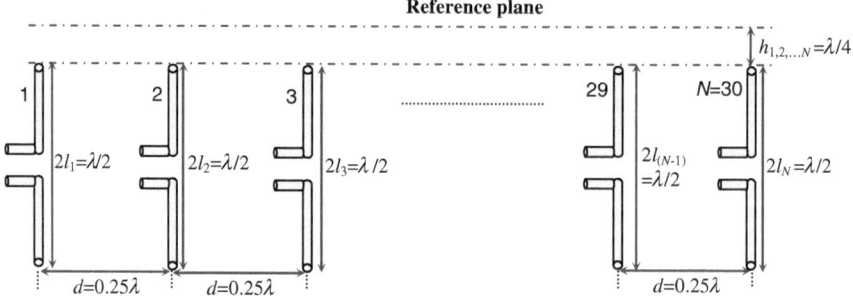

Fig. 7 An array of equal length $\lambda/2$ dipoles

3.1.2   Unequal-Length Dipole Array

Next, a center-fed dipole array in which the dipole lengths increment in steps of 0.01, starting from $\lambda/3$ is shown in Fig. 8a. Other configurations of dipole array are shown in Fig. 8b, c. The dipole lengths alter between $\lambda/4$ and $\lambda/3$, without and with an incrementing factor of 0.01, in Fig. 8b, c, respectively.

Next an array with random dipole lengths is shown in Fig. 8d. The random dipole lengths considered is given in Table 1. In all the cases, the staggered height of antenna element is taken as $\lambda/4$ below the reference plane. The results include the mutual coupling effect.

In Fig. 9, the RCS patterns of these configurations are shown and compared. All the design parameters of the array are taken to be same as that of Fig. 6. It can be observed from Fig. 9 that the level of lobes in the RCS pattern is maximum for an array shown in Fig. 8a, i.e., dipole array with dipole lengths $\lambda/3$ incrementing in steps of 0.01. Moreover, the level of RCS for the array with configuration of Fig. 8b is less as compared to the array, shown in Fig. 8c. However, the RCS is least for an array with random dipole length (Table 1). This comparison in RCS values at specular lobe for the above four configurations of center-fed dipole array with series feed network is given in Table 2.

In order to have clearer picture, the RCS pattern corresponding to the arrays with maximum (Fig. 8a) and minimum scattering (Fig. 8d) is shown as filled contour and contour plots in Figs. 10 and 11, respectively. The dB level of RCS lobes is indicated in terms of colors in filled contour plots; while in the contour plot, the RCS values (in dB) is indicated against the corresponding contour.

3.1.3   Role of Mutual Coupling Effect in RCS Estimation of Unequal Length Dipole Array

Next, an array of 20 thin-wire dipoles with alternative length of $\lambda/2$ and $\lambda/3$ is considered (Fig. 12) to analyze the effect of mutual coupling on the array RCS. The staggered height of dipoles w.r.t reference plane is considered to be $-\lambda/4$ for

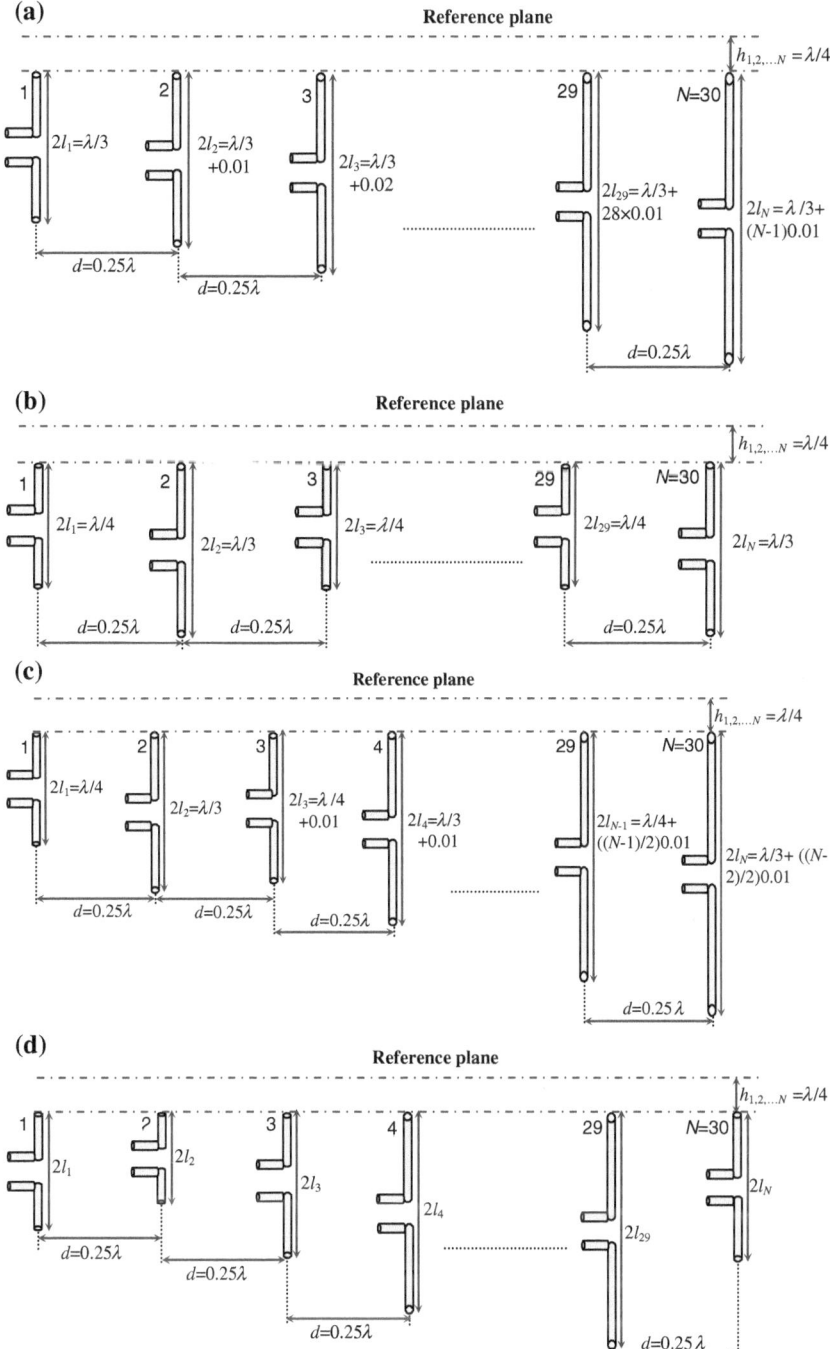

Fig. 8 **a** An array with dipole lengths $\lambda/3$ incrementing in steps of 0.01. **b** An array with dipole lengths $\lambda/4$ and $\lambda/3$ in odd and even positions. **c** ($\lambda/4$, $\lambda/3$) dipole array with the length incrementing in steps of 0.01. **d** Random length dipole array

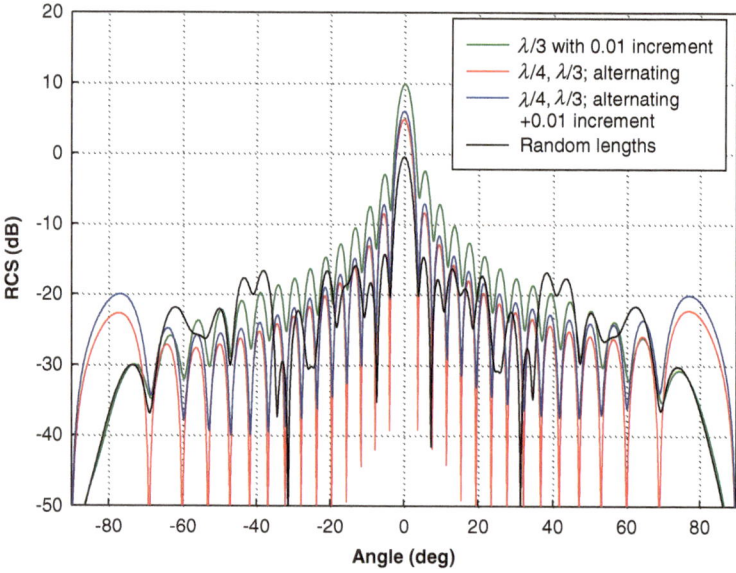

Fig. 9 Broadside RCS of unequal length series-fed dipole array in the presence of mutual coupling

Table 1 Lengths of dipoles in series-fed 30-element phased array

Dipole element	Dipole length (λ)	Dipole element	Dipole length (λ)
1	0.250	16	0.030
2	0.100	17	0.310
3	0.200	18	0.280
4	0.150	19	0.270
5	0.120	20	0.180
6	0.050	21	0.220
7	0.140	22	0.120
8	0.210	23	0.040
9	0.160	24	0.333
10	0.230	25	0.150
11	0.080	26	0.200
12	0.220	27	0.250
13	0.090	28	0.140
14	0.140	29	0.110
15	0.160	30	0.333

Table 2 RCS of series-fed dipole array with different configurations

S. no.	Array configuration	RCS level at specular lobe (dB)
1	$\lambda/3$ with 0.01 increment	9.7769
2	$\lambda/4$, $\lambda/3$; alternating	4.7682
3	$\lambda/4$, $\lambda/3$; alternating +0.01 increment	5.9395
4	Random lengths	−0.5145

odd-positioned elements and $\lambda/6$ for even–positioned elements. Other parameters are taken as $d = 0.2\lambda$, $Z_0 = 100 \, \Omega$, $Z_l = 40 \, \Omega$; with Taylor amplitude distribution (−45 dB SLL; $\bar{n} = 4$).

Figure 13 compares the RCS pattern of this array with and without mutual coupling effect. The results for broadside RCS ($\theta_s = 0°$) and scanned RCS (50°) are included.

It can be observed that the scanned RCS of the dipole array differ significantly for with and without mutual coupling cases. However the broadside RCS does not show variation in specular lobe RCS. This observation is similar to that in case of equal length dipole array (Sneha et al. 2012a, b). The variation in the scanned RCS pattern (with and without mutual coupling) further increases for larger values of scan angle. This may be due to the interelement interactions that vary the terminal impedances of the dipole elements and hence the reflections and RCS.

3.1.4 Effect of Terminal Impedance on RCS of Unequal-Length Dipole Array Including Mutual Coupling Effect

Next, the effect of varying the terminal impedances of the coupler ports on the RCS pattern of 25-element unequal length uniform ($d = 0.2\lambda$) dipole array is analyzed. The dipoles in the array are taken to be of lengths $\lambda/3$ and $\lambda/2$ at odd- and even positions of the array, respectively. The staggered heights of the elements w.r.t. reference plane are taken to be $\lambda/4$ and $-\lambda/3$ alternatively, shown in Fig. 14a.

The amplitude distribution is Dolph-Chebyshev with −40 dB side lobe level (SLL). The characteristic impedance of 75 Ω is considered. Figure 14b shows the broadside RCS pattern for the array when the terminating impedance is varied from 0 to 200 Ω. It is seen that the RCS levels decrease as the terminal impedance value increases from 0 to 35 Ω and further to 80 Ω. However for terminal impedance of 200 Ω, the RCS level shows rise.

This indicates that the decrease in RCS with variation of impedance terminating the coupler ports possess a limit. On reaching the limiting value of terminating impedance, the RCS level starts increasing with the load impedance Z_l. This effect is similar to the case of uniform equal length dipole arrays (Sneha et al. 2013). This feature can be exploited for RCS control of dipole array and its optimization.

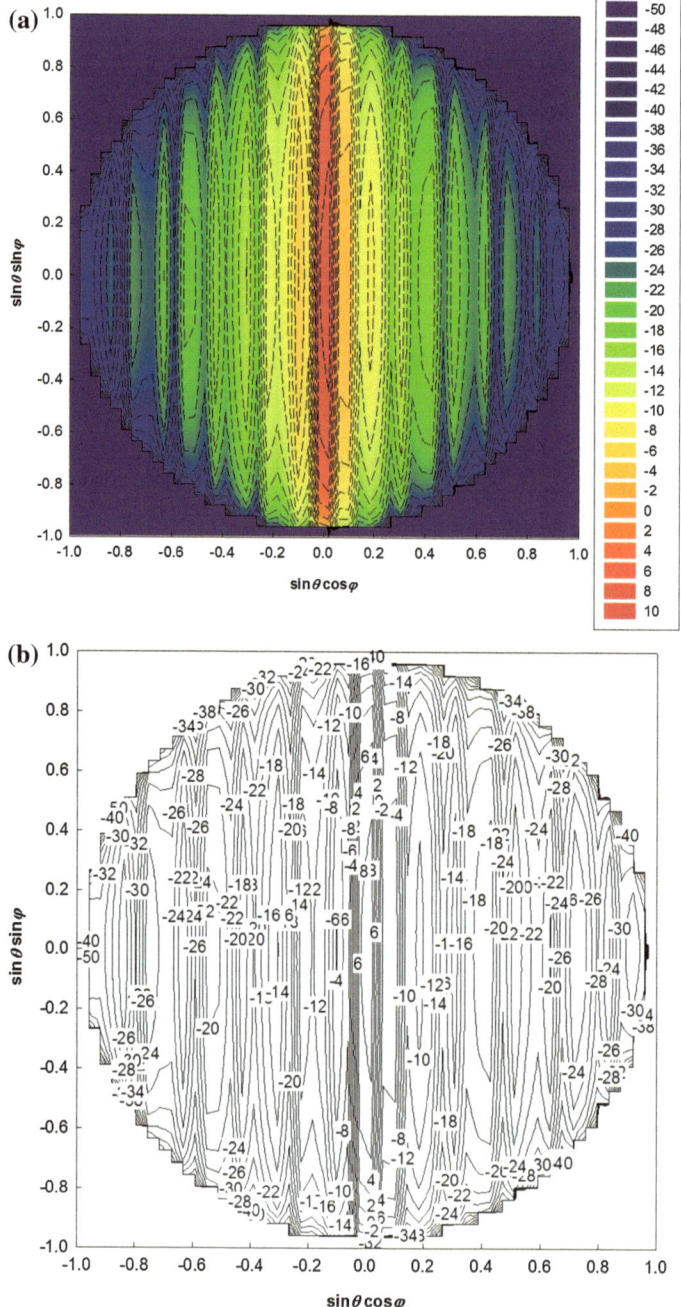

Fig. 10 Broadside array RCS of unequal length dipoles ($\lambda/3$; 0.01 increment). **a** Filled contour. **b** Contour

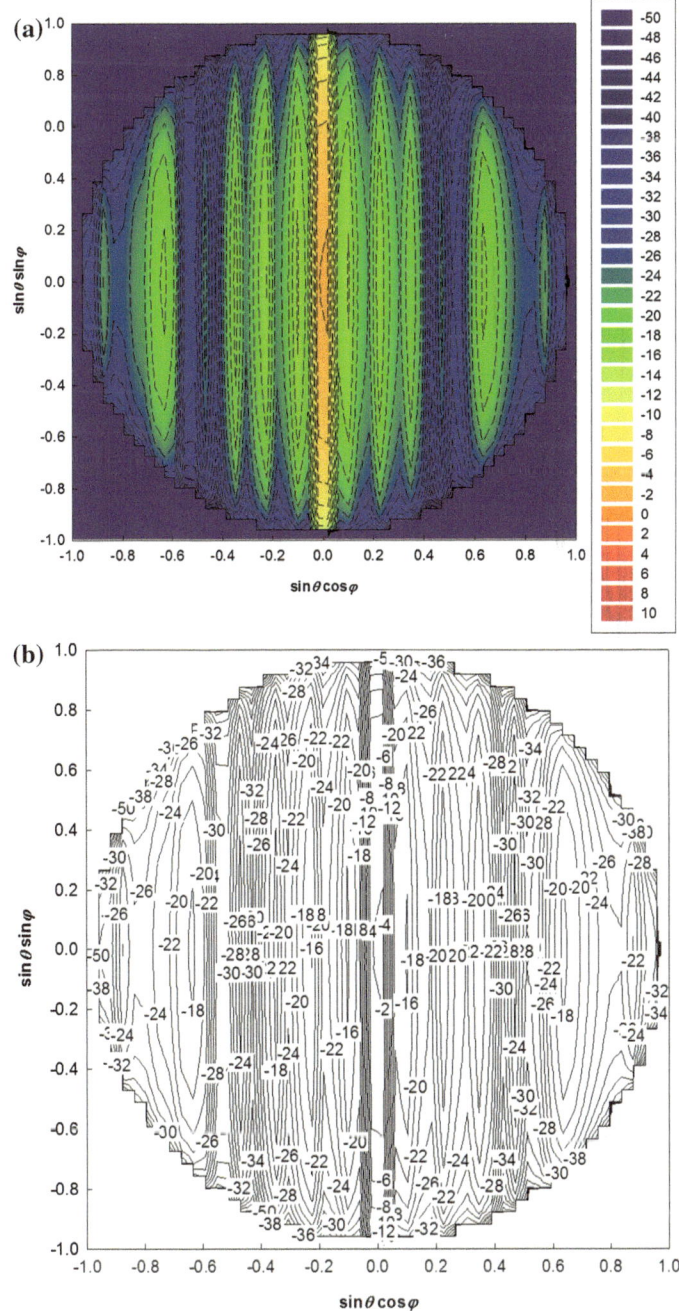

Fig. 11 Broadside array RCS of random length dipoles. **a** Filled contour. **b** Contour

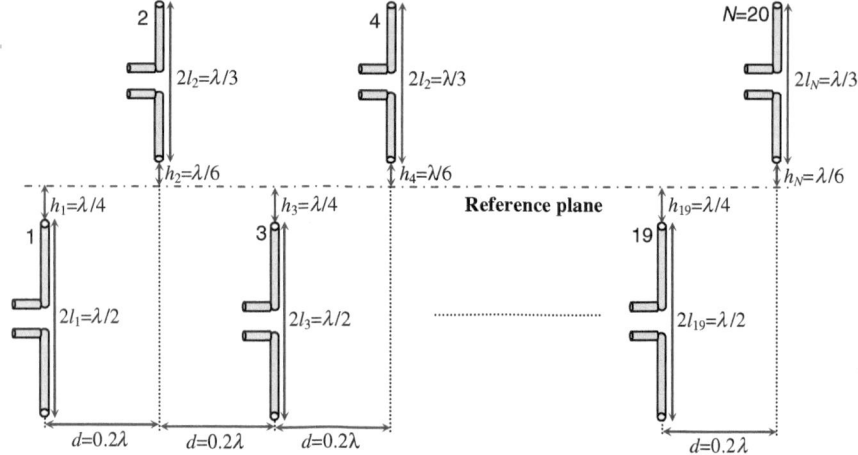

Fig. 12 An dipole array with alternative dipole lengths $\lambda/2$ and $\lambda/3$

3.2 RCS Estimation of Parallel-Fed Dipole Array

In this subsection, the RCS of parallel-fed linear dipole arrays with uniform spacing is analyzed.

3.2.1 Equal-Length Dipole Array

The geometry of 32-element dipole arrays is considered (Fig. 15). The staggered height of dipoles is $\lambda/4$ below the reference plane for odd-positioned dipoles. The computed broadside ($\theta_s = 0°$) and scanned ($\theta_s = 40°$) RCS patterns are shown in Fig. 16. The array parameters are $d = 0.3\lambda$, $Z_0 = 75\,\Omega$ and $Z_l = 20\,\Omega$; with cosine squared on a pedestal aperture distribution. The scattering till first level of couplers in the feed network is taken into account.

3.2.2 Unequal-Length Dipole Array

Next, the RCS of a 32-element unequal length linear dipole arrays is studied. Various configurations of unequal length dipole array with parallel feed are considered. Figure 17a shows an array with dipole lengths decrementing consistently in steps of 0.002, starting from $\lambda/2$. The dipoles at odd and even positions in the array are arranged at heights of $\lambda/4$ below and above the reference line, respectively.

Another array with the dipoles of length $\lambda/2$ and $\lambda/3$, in alternate positions is shown by Fig. 17b. Here the dipole length decreases in steps of 0.002 as one move along the array. Figure 17c shows the third array with random dipole lengths.

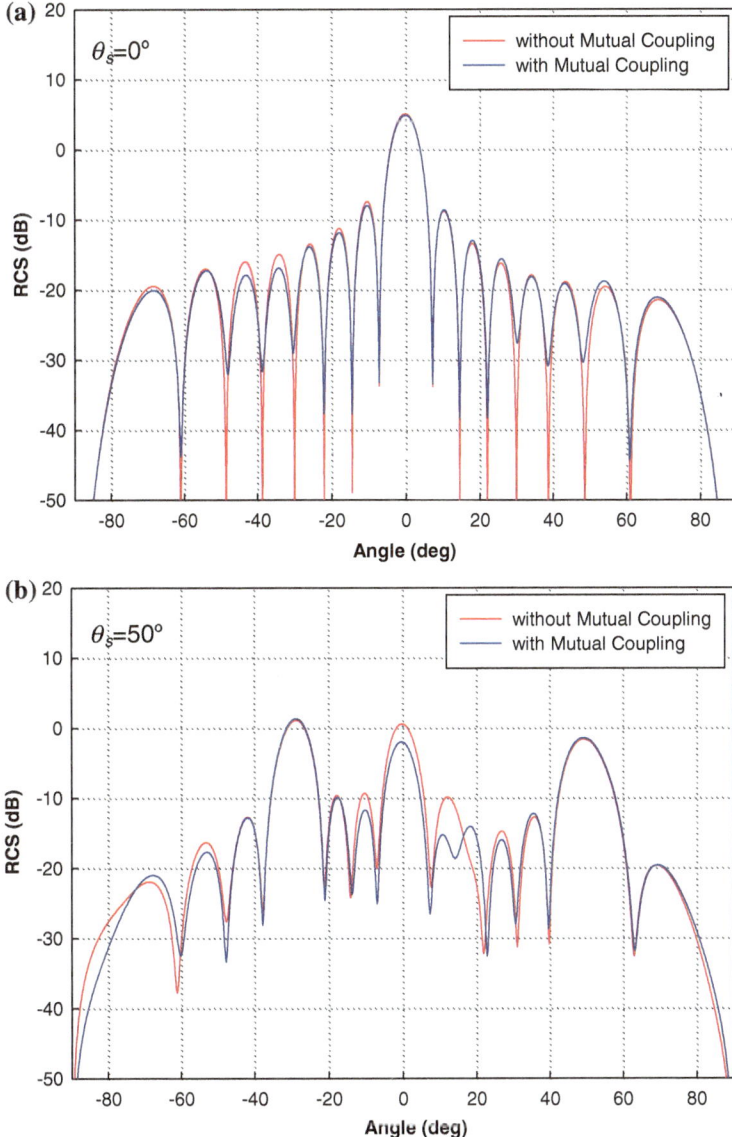

Fig. 13 RCS patterns of 20-element unequal length linear dipole array, with and without mutual coupling effect. **a** $\theta_s = 0°$. **b** $\theta_s = 50°$

The dipole lengths taken are given in Table 3. The corresponding RCS patterns of these dipole arrays including mutual coupling effect are compared in Fig. 18. The dipoles are arranged with an interelement spacing of 0.4λ. The amplitude distribution considered is Taylor distribution (−45 dB SLL; $\bar{n} = 4$). The characteristic impedance and load impedance are taken as 75 and 20 Ω, respectively.

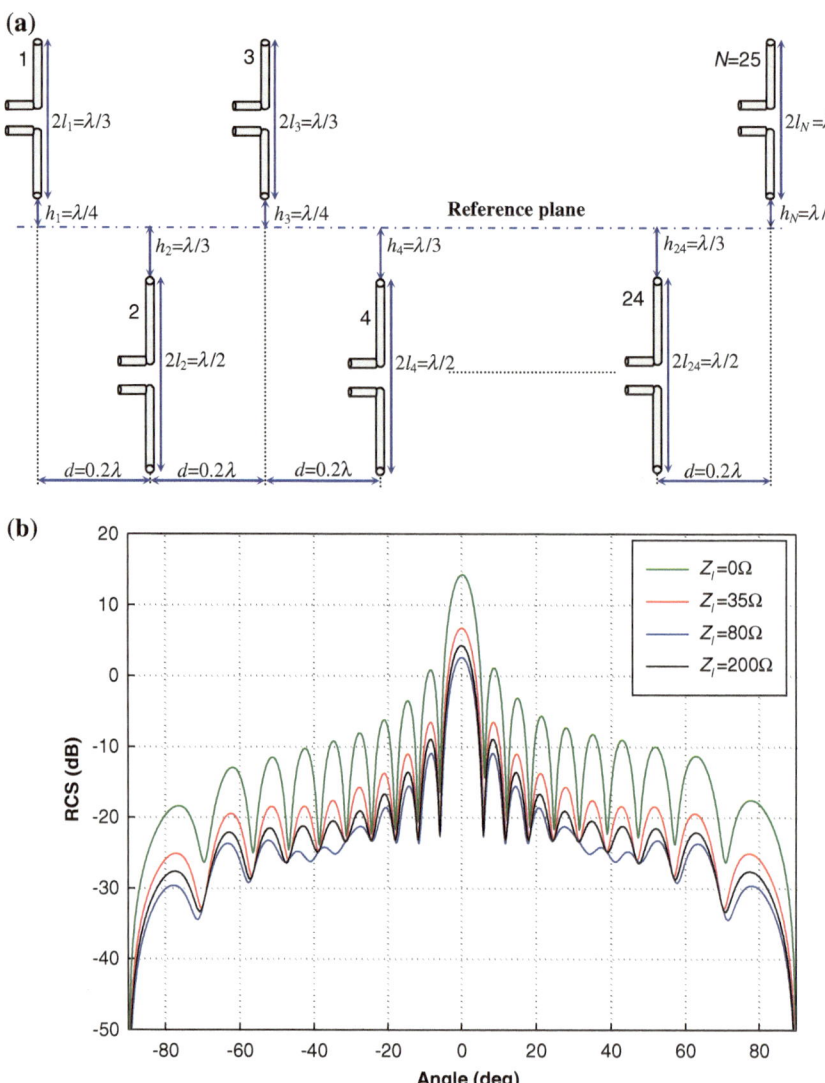

Fig. 14 **a** A dipole array with dipole lengths alternating between $\lambda/3$ and $\lambda/2$. **b** RCS pattern of 25-element unequal length linear dipole array in the presence of mutual coupling for varying terminal impedances

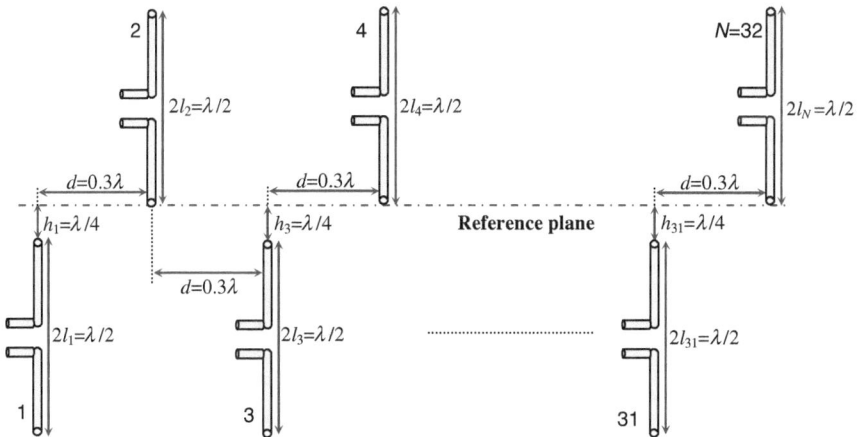

Fig. 15 A dipole array of equal length $\lambda/2$ dipoles

It is apparent that the RCS of array is lowest at both specular lobe ($\theta = 0°$) and I level coupler mismatch lobes ($\theta = \pm38°$) for random dipole lengths (Fig. 17c). However the levels of RCS at the lobes arising due to the mismatches at II level of couplers ($\theta = \pm17°$) in the feed network show a slight increase. On the other hand, an array with configuration shown in Fig. 17b shows reduced specular level but increased I level coupler mismatch lobe level, as compared to the dipole array in Fig. 17a. However, this array does not have any noticeable lobes which arise due to the impedance mismatches at coupler level II of the parallel feed network. This indicates that the dipole length is an important parameter that can be optimized toward low observable platform. The RCS values at specular and lobes due to mismatches at first coupler level for the above three configurations of center-fed dipole array with parallel feed network is given in Table 4.

3.2.3 Mutual Coupling Effect in Unequal-Length Dipole Array with Parallel Feed Network

For demonstrating the role of mutual coupling in unequal length dipole array with parallel feed network, a dipole array is considered with odd-positioned dipole lengths incrementing consistently by 0.01, starting from $\lambda/3$. Further, the dipole lengths at the even positions decrement consistently in the steps of 0.01 starting from $\lambda/2$. All the dipole elements are at the height of $\lambda/4$ above the reference plane (Fig. 19). Figure 20 shows the broadside RCS pattern of a 16-element dipole array in which the elements are arranged as per Fig. 19. The pattern is compared for with and without mutual coupling effect.

The interelement spacing is taken as 0.4λ, while the characteristic impedance and the terminating impedances are assumed to be 75 and 200 Ω, respectively.

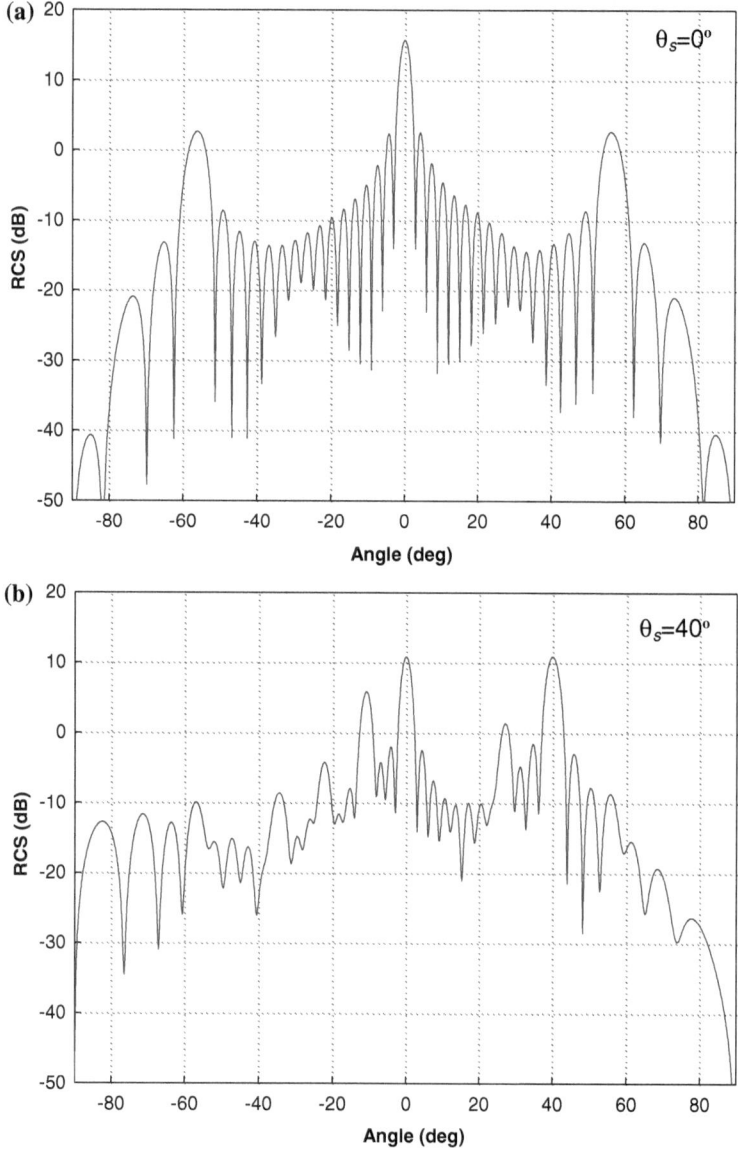

Fig. 16 RCS pattern of 32-element equal length ($\lambda/2$) linear dipole array in the presence of mutual coupling. **a** $\theta_s = 0°$. **b** $\theta_s = 40°$

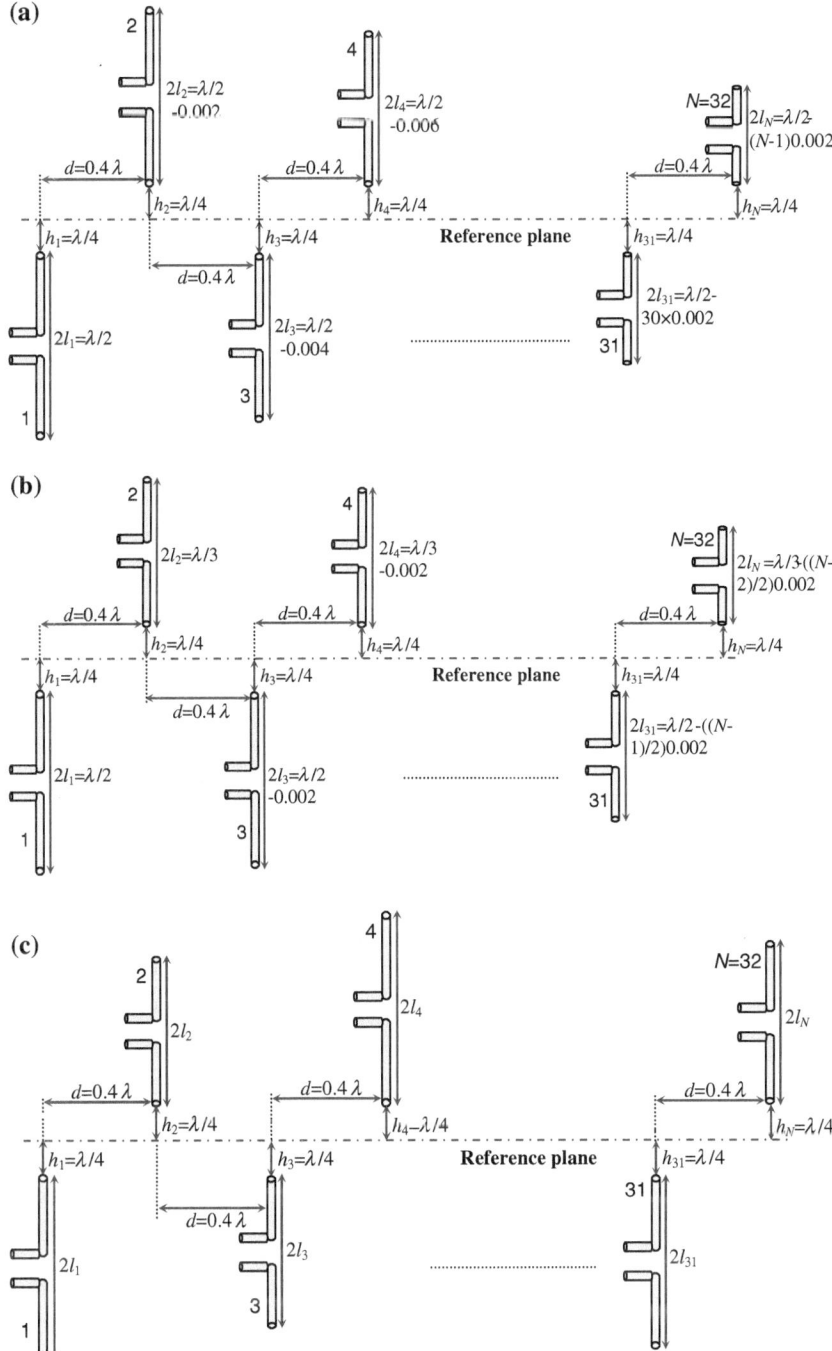

Fig. 17 **a** An array of 32-element dipoles of length $\lambda/2$ decrementing consistently by 0.002. **b** An array with dipoles lengths $\lambda/2$ and $\lambda/3$, decrementing consistently by 0.002. **c** An array with random length dipoles

Table 3 Dipole length in a parallel-fed 32-element phased array

Dipole element	Dipole length (λ)	Dipole element	Dipole length (λ)
1	0.450	17	0.500
2	0.250	18	0.360
3	0.425	19	0.333
4	0.500	20	0.425
5	0.300	21	0.397
6	0.400	22	0.298
7	0.390	23	0.350
8	0.340	24	0.420
9	0.444	25	0.345
10	0.500	26	0.375
11	0.388	27	0.500
12	0.241	28	0.256
13	0.410	29	0.428
14	0.370	30	0.399
15	0.430	31	0.433
16	0.490	32	0.285

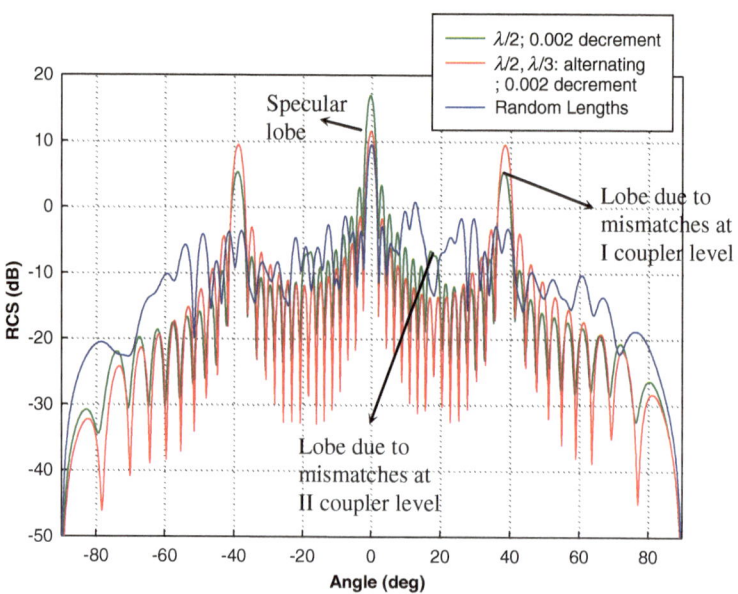

Fig. 18 Broadside RCS patterns of 32-element unequal length linear dipole array with parallel feed network

Table 4 RCS of parallel-fed dipole array with different configurations

S. no.	Array configuration	RCS level at specular lobe (dB)	RCS level at lobes due to coupler level mismatches (dB)	
			I Level	II Level
1	$\lambda/2$ with 0.002 decrement	16.6092	5.2465	−6.6776
2	$\lambda/2$, $\lambda/3$; alternating +0.002 increment	11.4410	9.3096	−13.8024
3	Random lengths	9.4614	−6.3484	−12.1510

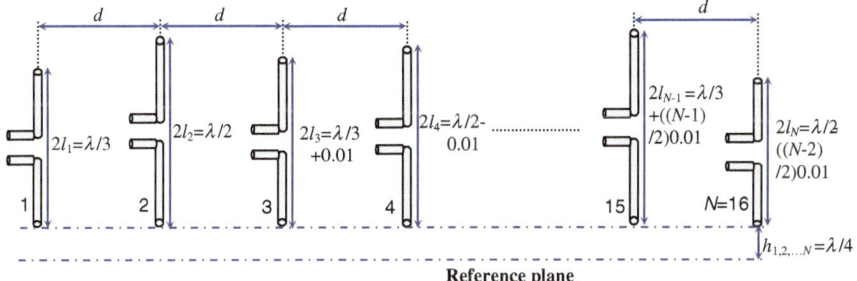

Fig. 19 Unequal length phased array; odd-positioned dipoles with $\lambda/3$ dipole lengths incrementing at 0.01 step while even-positioned dipoles with $\lambda/2$ dipole length decrementing at 0.01 step

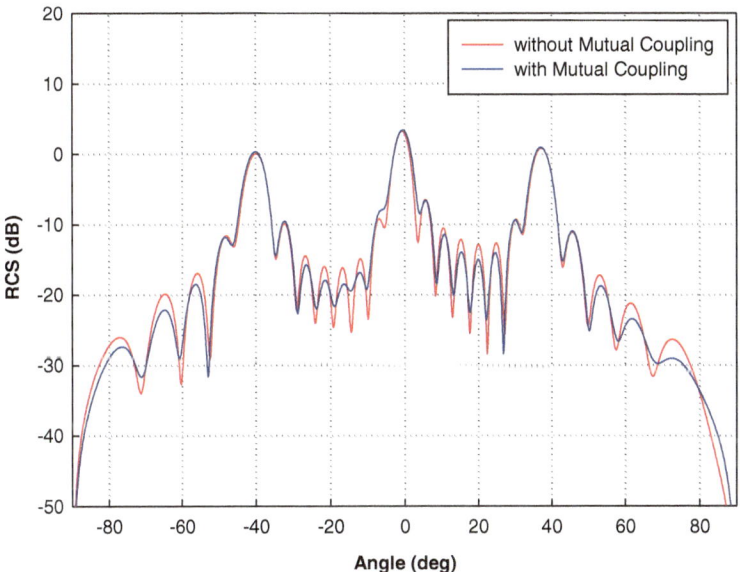

Fig. 20 Comparison of broadside RCS pattern of 16-element unequal length linear parallel-fed dipole array for with and without mutual coupling effect; $d = 0.4\lambda$

The amplitude distribution exciting the array is cosine squared on a pedestal. The scattering till first level of couplers is taken into account. It can be observed that the mutual coupling effect alters the level of lobes in the RCS pattern, which might be due to the variation in terminal impedances of dipoles.

3.2.4 Role of Terminating Impedance on RCS of Unequal-Length

As discussed earlier, terminating impedance plays an important role in RCS of dipole array. In order to check whether this is true in case of parallel feed as well, a 64-element array (Fig. 21) is considered, in which the dipoles of lengths $\lambda/3$ and $\lambda/2$ are arranged alternately with staggered heights of $\lambda/4$ below and above the reference plane, respectively. The spacing between elements is 0.4λ. The influence of impedance terminating the coupler ports, on the RCS pattern of such an unequal length parallel-in-echelon dipole array is shown in Fig. 22. Dipoles are excited by uniform unit amplitude distribution and the characteristic impedance is taken as 75 Ω. It is observed that the RCS level is maximum for 0 Ω termination and decreases as the impedance is increased to 25 Ω and further to 50 Ω.

However, for 160 Ω termination, the level of RCS lobes is seen to increase. This observation is similar to the case of equal length dipole array with parallel feed (Sneha et al. 2013).

In order to further emphasize this fact, another array of 32 unequal length dipoles is considered. The dipoles are assumed to be of alternating lengths $\lambda/2$ and $\lambda/3$ (Fig. 23). The spacing between the array elements, excited by Dolph-Chebyshev amplitude distribution (-40 dB SLL) is taken as 0.3λ. The characteristic impedance is 75 Ω while the terminating impedance is varied. The RCS levels are shown as filled contour and contour plots in Figs. 24, 25, and 26. The results are shown for

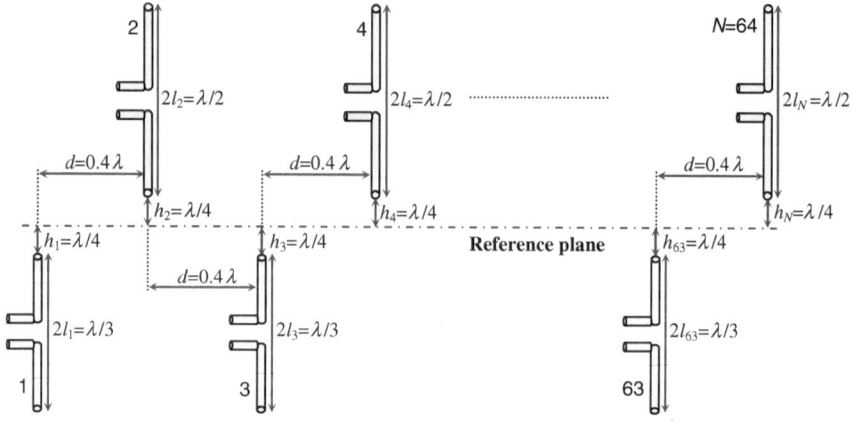

Fig. 21 A parallel-in-echelon array with dipoles of length $\lambda/3$ and $\lambda/2$ at alternate positions

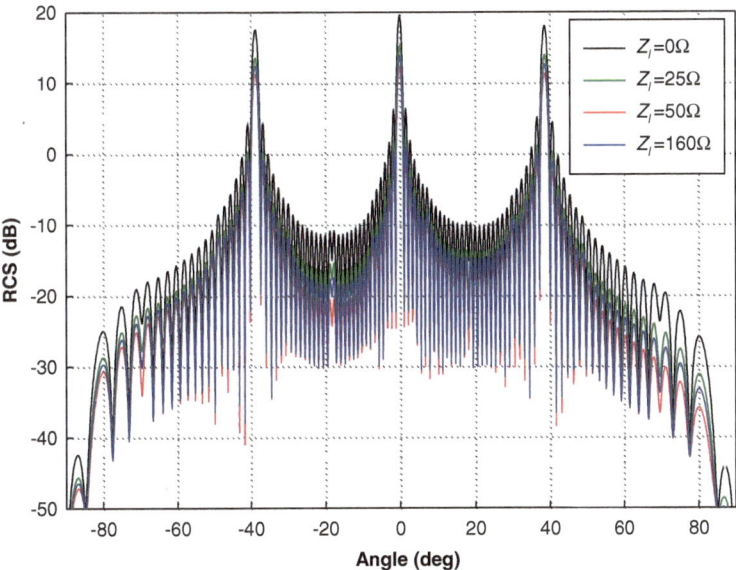

Fig. 22 Effect of varying the terminating impedances on the broadside RCS pattern of 64-element unequal length parallel-in-echelon dipole array with parallel feed network

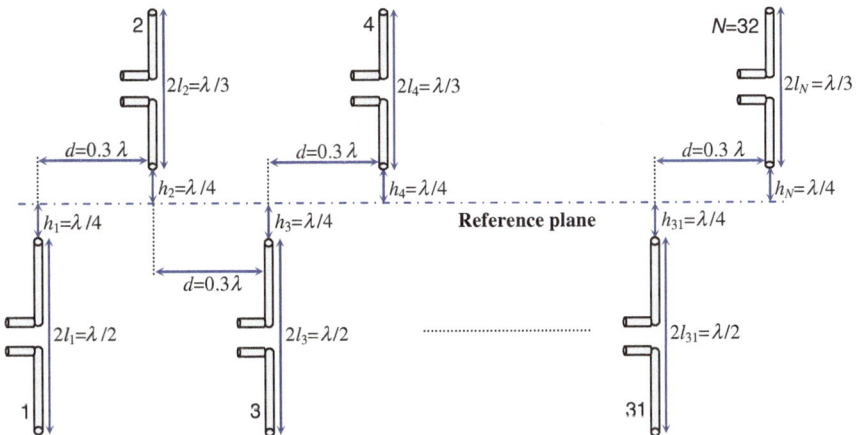

Fig. 23 An array with dipoles lengths $\lambda/2$ and $\lambda/3$ in odd- and even positions, respectively

the terminating impedances of 30, 100, and 200 Ω in Figs. 24, 25, and 26, respectively. It can be observed that the level of RCS, indicated by color in case of filled contour and as dB value in contour plots, decreases as the impedance value is increased to 100 from 30 Ω.

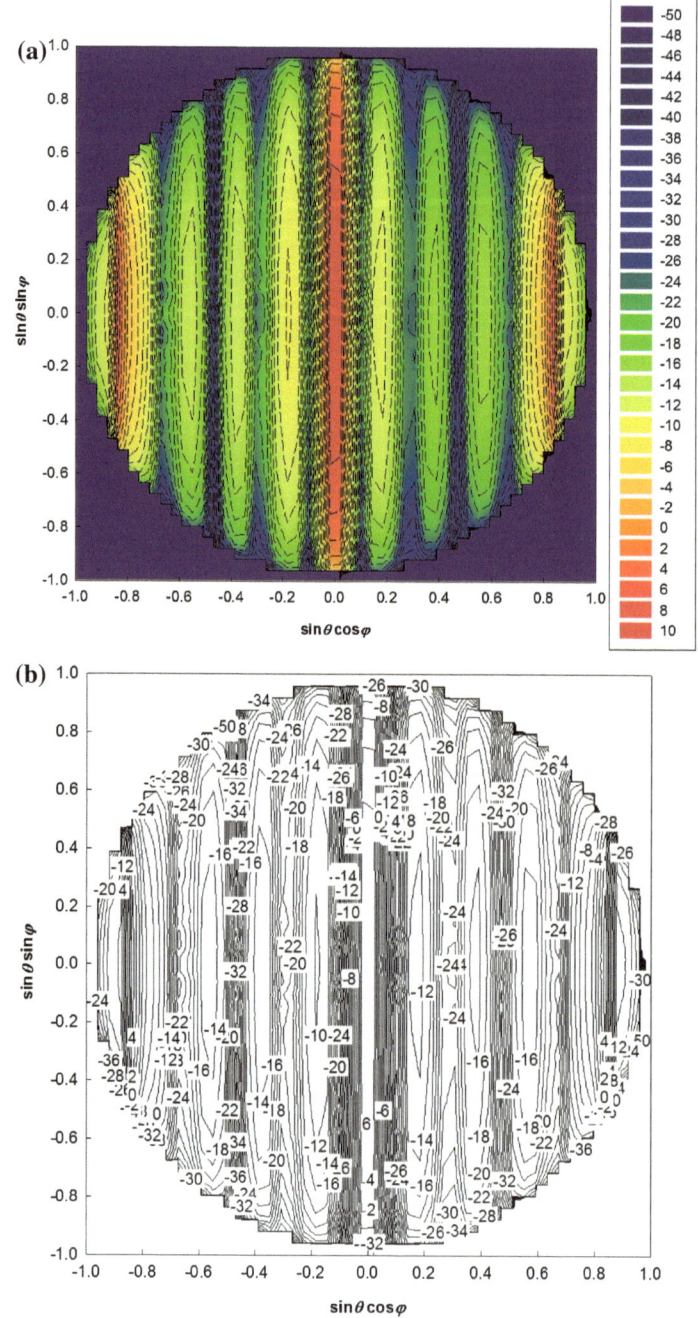

Fig. 24 Broadside array RCS of parallel-fed unequal length dipole array terminated by 30 Ω.
a Filled contour. **b** Contour

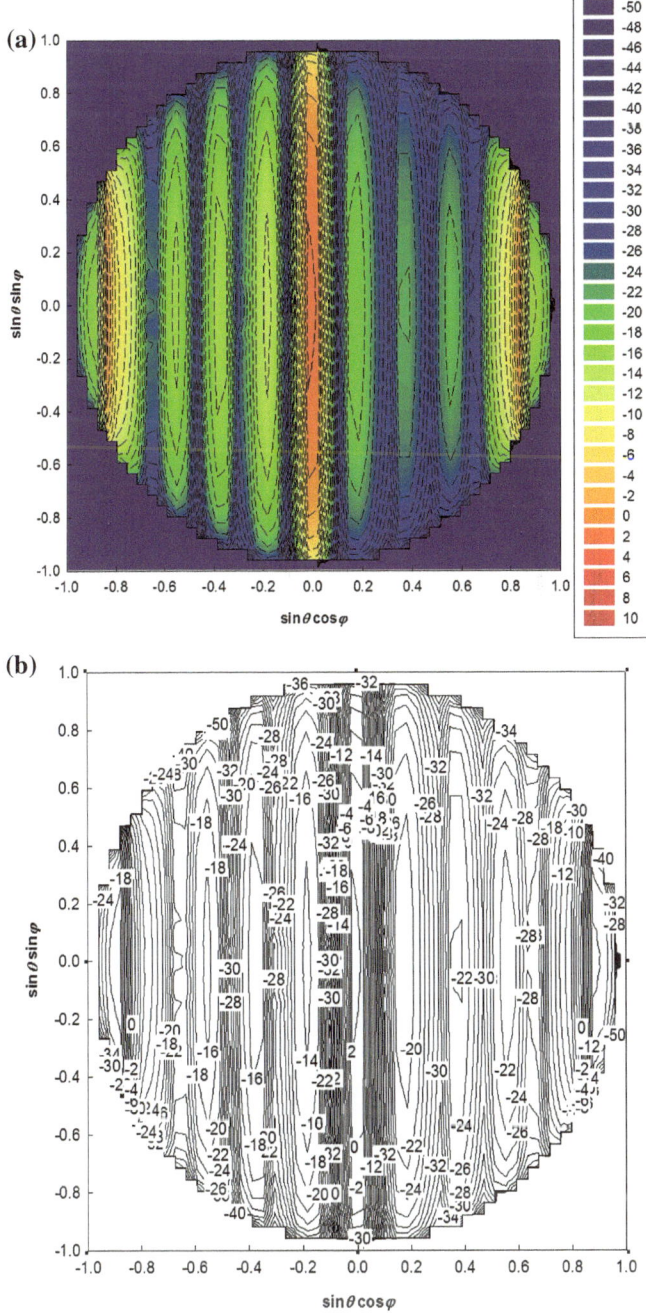

Fig. 25 Broadside array RCS of parallel-fed unequal length dipoles terminated by 100 Ω. **a** Filled contour. **b** Contour

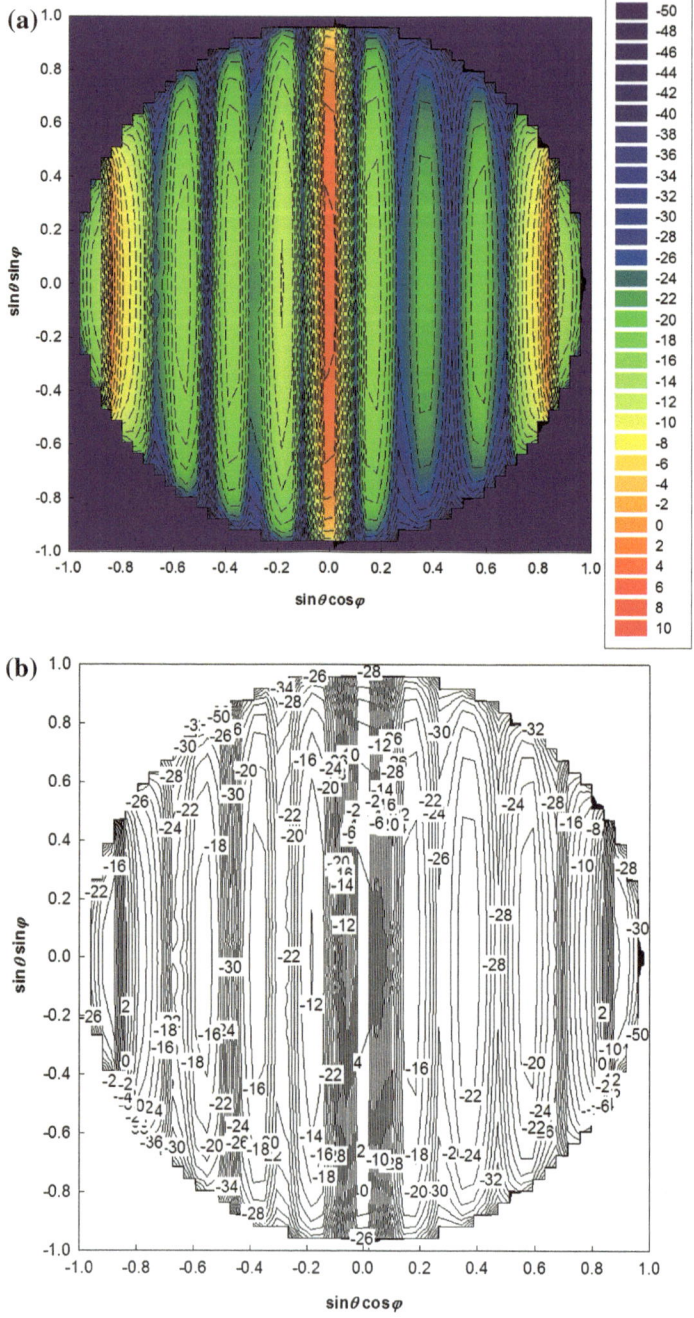

Fig. 26 Broadside array RCS of parallel-fed unequal length dipoles terminated by 200 Ω. **a** Filled contour. **b** Contour

However when the impedance value is increased further to 200 Ω, the level of RCS increases. This indicates that the concept of limiting impedance holds good, irrespective of the design parameters including length of dipoles, scan angle, amplitude distribution, and type of feed network. Thus optimization of the impedances used to terminate the coupler ports has a potential toward the RCS reduction of a phased array.

4 Conclusion

In this book, the RCS estimation of unequal length linear dipole arrays with uniform spacing is presented. The formulation for total RCS of dipole array with different combination of unequal lengths is done. The computations are valid for any arbitrary array configurations, including side-by-side arrangement, parallel-in-echelon, etc. However for the skewed dipole array configuration, the expressions need modifications. The reflections occurring within the antenna system and the effects of mutual coupling between the array elements are considered to arrive at the total RCS corresponding to two types of feed networks viz. series feed and parallel feed. The scattering due to higher order reflections is neglected for both types of feeds and the computations are restricted till second level of couplers in parallel feed network. The expressions presented include the dependence of RCS of dipole array on design parameters viz. dipole length, interelement spacing, geometrical and electrical properties of couplers and terminal impedances.

The variation in the length of dipole elements is shown to affect the RCS pattern of the array significantly. It is shown that the length of array elements would be a potential parameter for optimization to obtain low observable targets. Moreover, the mutual coupling affects the RCS pattern of an unequal length dipole array for both types of feeds. The variation in RCS pattern becomes further noticeable as the scan angle of array increases, irrespective of any other design criteria. The terminating impedance is another important parameter that can be exploited for RCS control. There is a limiting value of terminating impedance beyond which RCS value of phased array increases. This is true for either type of feed networks. In broad sense, the effect of varying the design parameters on RCS pattern is similar in both equal- and unequal length dipole arrays.

References

Balanis, C.A. 2005. *Antenna theory, analysis and design*, 1117 p. New Jersey: Wiley. ISBN:0-471-66782-X.

Jenn, D.C. 1995. *Radar and laser cross section engineering*. Washington, DC: AIAA Education Series, 476 p. ISBN:1-56347-105-1.

Jenn, D.C., and S. Lee. 1995. In-band scattering from arrays with series feed networks. *IEEE Transactions on Antennas and Propagation* 43: 867–873.

Jenn, D.C., and V. Flokas. 1996. In-band scattering from arrays with parallel feed networks. *IEEE Transactions on Antennas and Propagation* 44: 172–178.

King, H.E. 1957. Mutual impedance of unequal-length antennas in echelon. *IRE Transactions on Antennas and Propagation* 5: 306–313.

Liao, Y., S. Yang, H. Ma and Y. Hou. 25–28 June 2006a. Research on characteristics of finite dipole array. *Proceeding of IEEE International Conference on Communications, Circuits and Systems*, Guilin, China, vol. 2, pp. 929–932, 25–28.

Liao, Y., S. Yang, H. Ma and Y. Hou. 21–23 June 2006b. Research on scattering property of finite dipole array. *Proceedings of 6th International Conference on ITS Telecommunications*, Chengdu, China, pp. 412–415.

Liu, Y., and L. You. 1–3 November 2011. Research on the estimation and reduction measures of antenna mode RCS of airborne phased array. *IEEE 4th International Symposium on Microwave, Antenna Propagation and EMC Technologies for Wireless Communications (MAPE)*, Beijing, China, pp. 79–82.

Niow, C.H., Y.T. Yu, and H.T. Hui. 2011. Compensate for the coupled radiation patterns of compact transmitting antenna arrays. *IET Microwaves and Antennas Propagation* 5: 699–704.

Sneha, H.L., Hema Singh, and R.M. Jha. June 2012a. *Radar cross section (RCS) of a series-fed dipole array including mutual coupling effect*. CSIR-National Aerospace Laboratories, Bangalore, India, Project Document PD AL 1222, 36 p.

Sneha, H.L., Hema Singh, and R.M. Jha. August 2012b. Mutual coupling effects for radar cross section (RCS) of a series-fed dipole antenna array. *International Journal of Antennas and Propagation* 2012:20 p.

Sneha, H.L., Hema Singh, and R.M. Jha. January 2013. *Back-scattering cross section of a parallel-fed dipole array including mutual coupling effect*. CSIR-National Aerospace Laboratories, Bangalore, India, Project Document PD CEM 1306, 51 p.

Zengrui, L., and W. Junhong. 26-29 October 2006. Study on the scattering property of the impedance terminated dipole array by FDTD Method. *Proceedings of 7th International Symposium on Antennas Propagation and EM Theory* (ISAPE 2006), Guilin, China, pp. 1065–1068.

Zengrui, L., W. Junhong, L. Limei and Z. Xueqin. 16–17 August 2007. Study on the scattering property of the impedance terminated dipole array with finite reflector by FDTD method. *Proceedings of IEEE International Symposium on Microwave, Antenna, Propagation and EMC Technologies for Wireless Communications*, Hangzhou, China, pp. 1003–1007.

Zhang, S., S.X. Gong, Y. Guan, J. Ling, and B. Lu. 2010. A new approach for synthesizing both the radiation and scattering patterns of linear dipole antenna array. *Journal of Electromagnetic Waves and Application* 24: 861–870.

About the Book

This book presents a detailed and systematic analytical treatment of scattering by an arbitrary dipole array configuration with unequal-length dipoles, different inter-element spacing and load impedance. It provides a physical interpretation of the scattering phenomena within the phased array system. The antenna radar cross section (RCS) depends on the field scattered by the antenna towards the receiver. It has two components, viz. structural RCS and antenna mode RCS. The latter component dominates the former, especially if the antenna is mounted on a low observable platform. The reduction in the scattering due to the presence of antennas on the surface is one of the concerns towards stealth technology. In order to achieve this objective, a detailed and accurate analysis of antenna mode scattering is required. In practical phased array, one cannot ignore the finite dimensions of antenna elements, coupling effect and the role of feed network while estimating the antenna RCS. This book presents the RCS estimation of an array with unequal-length dipoles. The signal reflections within the antenna system and the mutual coupling effect are considered to arrive at the total RCS for series and parallel feed. The computations are valid for any arbitrary array configurations, including side-by-side arrangement, parallel-in-echelon, etc.

© The Author(s) 2016   39
H. Singh et al., *Scattering Cross Section of Unequal Length Dipole Arrays*,
SpringerBriefs in Computational Electromagnetics,
DOI 10.1007/978-981-287-790-1

Author Index

© The Author(s) 2016
H. Singh et al., *Scattering Cross Section of Unequal Length Dipole Arrays*,
SpringerBriefs in Computational Electromagnetics,
DOI 10.1007/978-981-287-790-1

Subject Index

© The Author(s) 2016
H. Singh et al., *Scattering Cross Section of Unequal Length Dipole Arrays*,
SpringerBriefs in Computational Electromagnetics,
DOI 10.1007/978-981-287-790-1

43